# SPECIFICATION WRITING AND MANAGEMENT

# QUALITY AND RELIABILITY

*A Series Edited by*

**Edward G. Schilling**

Center for Quality and Applied Statistics
Rochester Institute of Technology
Rochester, New York

# SPECIFICATION WRITING AND MANAGEMENT

## Max McRobb
### *Bedford, England*

## CRC Press
Taylor & Francis Group
Boca Raton  London  New York

CRC Press is an imprint of the
Taylor & Francis Group, an **informa** business

First published 1989 by Marcel Dekker, Inc

Published 2019 by CRC Press
Taylor & Francis Group
6000 Broken Sound Parkway NW, Suite 300
Boca Raton, FL 33487-2742

© 1989 by Taylor & Francis Group, LLC
CRC Press is an imprint of Taylor & Francis Group, an Informa business

First issued in paperback 2019

No claim to original U.S. Government works

ISBN-13: 978-0-367-45100-4 (pbk)
ISBN-13: 978-0-8247-8082-1 (hbk)

Visit the Taylor & Francis Web site at
http://www.taylorandfrancis.com

and the CRC Press Web site at
http://www.crcpress.com

Library of Congress Cataloging-in-Publication Data

McRobb, Max
    Specification writing and management / Max McRobb.
        p. cm—(Quality and reliability)
    Includes index.
    ISBN 0-8247-8082-5 (alk. paper)
    1. Quality control. 2. Specification writing. I. Title.
II. Series.
TS156.M39    1989
658.5'62—dc20                    89-31388
                                    CIP

# About the Series

The genesis of modern methods of quality and reliability will be found in a simple memo dated May 16, 1924, in which Walter A. Shewhart proposed the control chart for the analysis of inspection data. This led to a broadening of the concept of inspection from emphasis on detection and correction of defective material to control of quality through analysis and prevention of quality problems. Subsequent concern for product performance in the hands of the user stimulated development of the systems and techniques of reliability. Emphasis on the consumer as the ultimate judge of quality serves as the catalyst to bring about the integration of the methodology of quality with that of reliability. Thus, the innovations that came out of the control chart spawned a philosophy of control of quality and reliability that has come to include not only the methodology of the statistical sciences and engineering, but also the use of appropriate management methods together with various motivational procedures in a concerted effort dedicated to quality improvement.

This series is intended to provide a vehicle to foster interaction of the elements of the modern approach to quality, including statistical applications, quality and reliability engineering, management, and motivational aspects. It is a forum in which the subject matter of these various areas can be brought together to allow for effective integration of appropriate techniques. This will promote the true benefit of each, which can be achieved only through their interaction. In this sense, the whole of quality and reliability is greater than the sum of its parts, as each element augments the others.

The contributors to this series have been encouraged to discuss fundamental concepts as well as methodology, technology, and procedures at the leading edge of the discipline. Thus, new concepts are placed in proper perspective in these evolving disciplines. The series is intended for those in manufacturing, engineering, and marketing and management, as well as the consuming public, all of whom have an interest and stake in the improvement and maintenance of quality and reliability in the products and services that are the lifeblood of the economic system.

The modern approach to quality and reliability concerns excellence: excellence when the product is designed, excellence when the product is made, excellence as the product is used, and excellence throughout its lifetime. But excellence does not result without effort, and products and services of superior quality and reliability require an appropriate combination of statistical, engineering, management, and motivational effort. This effort can be directed for maximum benefit only in light of timely knowledge of approaches and methods that have been developed and are available in these areas of expertise. Within the volumes of this series, the reader will find the means to create, control, correct, and improve quality and reliability in ways that are cost effective, that enhance productivity, and that create a motivational atmosphere that is harmonious and constructive. It is dedicated to that end and to the readers whose study of quality and reliability will lead to greater understanding of their products, their processes, their workplaces, and themselves.

*Edward G. Schilling*

# Foreword

I have had the pleasure of knowing Max McRobb personally for many years. He has always been an enthusiast for the principles of quality assurance, especially for the part that well-written specifications play in defining the product quality that has to be attained.

Any work on specification writing and management is of value to all standards bodies and to the organizations that support their work, quite apart from all those who write the specifications used in commerce and industry.

There are relatively few works dealing with specification writing. In this book the author has conveyed his findings with enthusiasm and clarity. Not everyone will agree with all he says. Not all he says is written in the same terms as British Standards 0-A standard for standards. But because he has scrupulously avoided jargon and has taken pains to put himself in the position of the reader, no reader is likely to be left in doubt about the message he conveys.

The book will be compulsive reading for anyone concerned with the preparation of standards in a world that is shrinking due to improvements in communications and where many specifications have to hold their clarity when translated into other languages. It is also a world which can no longer afford the cost of poorly written specifications.

*Specification Writing and Management* is therefore a valuable contribution in a neglected field. The inadequacies of specifications in many areas of the economy need the attention of management drawn to them. This publication is particularly timely in that it coincides with the Department of Trade and Industry's Enterprise Initiative.

I feel privileged to write the foreword to what should become a valuable reference to all those who have the unglamorous but most essential task of producing specifications that define accurately the market needs and thus improve the profitability of their companies.

*Rear Admiral D. G. Spickernell, CB*
*Former Director General, British Standards Institution,*
*and Vice President, International Standards Organisation*

# Preface

My experience over many years has convinced me of the fact that many of the problems that arise in industry originate with the specifications—specifications for which inadequate care has been expended on their preparation.*

The low level of the quality of some specifications produces a similar low regard for them on the part of users. This often results in considerably increased costs—the lost costs. Throughout the book the importance of these lost costs is stressed. The cause of these costs is usually unsuspected, certainly not attributed to the specifications, and is normally hidden by being charged to other causes, for example, faulty processes. The extent of these costs is not known but it has been suggested that for British industry they could amount to no less than $200 million a year.

On a number of occasions over the years I have had opportunities to recognize this problem in practice and have been able to develop local solutions. Complete solutions are not really possible until the problem is recognized as being industry-wide, which, undoubtedly, it is. Unfortunately, it is not easy to persuade industry

---

*For the purposes of this book the term *specification* should be taken to include *standards, operational procedures*, and, indeed, any form of *written instruction*. They are all, in the sense that they state requirements, specifications.

that the problem exists. During the last 15 years I have often had an opportunity to discuss specification problems with people in a number of industries in a number of countries. It was clear from these discussions that the problem is by no means confined to the United Kingdom; it is worldwide.

One outcome of these discussions was the development of a short course that deals extensively with specification writing and management. This course has been presented in a number of countries and has been well received. As a result, I was encouraged to expand considerably the course material into this book. Although a number of official as well as private publications have appeared that deal with the subject of the preparation of specifications, most of them do not deal with the fundamentals of specification writing. They are concerned primarily with the nature of the technical content. Those that do touch on this subject, for the most part, do so almost in passing.

As far as I am aware, this is the first book ever written to deal solely with the fundamentals of specification writing and management. It is thoroughly practical since it is based on the real experience of more than 20 years. There are no hypothetical cases. Every example discussed, whether of specification or situation, is one that I have experienced. Each case study detailed was a real situation. Each of the solutions for these situations is the actual one developed for the particular case, though it has become clear to me that the same problems arise in industry everywhere. It has been claimed by an American consultant to the United Nations that the faults of the developed nations in specifications are being repeated by the developing nations—faults produced by the national standards bodies of the developed nations.

I do not claim that the solutions described in this book can be taken as written and applied directly to situations faced by readers. But I maintain that from the information given, the examples, and the real industrial situations described, readers will be able to adapt these solutions or devise suitable alternative solutions for their particular needs, whether they work for large or small companies or organizations.

I hope that the improvement in the quality of specifications that will result from the use of this book will have a number of benefits: an increase in the regard for specifications in industry, a considerable reduction in the number of problems, and a reduction in lost costs, which, in turn, will help industry to become more profitable.

There are a number of acknowledgments that I wish to make. First, to the British Standards Institution, for permission to make short extracts from a number of their publications and to make reference to others; to Bill Windebank, for reading an early draft and providing numerous suggestions for improvements to the text; to Alan MacGregor, my last boss, for reading the final draft at short notice and providing encouragement; and finally, to my wife, who made laborious word-by-word checks for spelling and typographical errors. Perhaps gratitude should also be expressed to the host of unknown writers of specifications over the past quarter of a century who have provided such a feast of examples that made this book possible.

*Max McRobb*

# Contents

# 1

# A Brief Historical Review

The following quotation regarding Solomon's temple is from Kings 1:6:

*In the inner sanctuary he made two cherubim of olive wood, each ten cubits high.*

*Five cubits was the length of one wing of the cherub, and five cubits was the length of the other wing of the cherub.*

*It was ten cubits from the tip on one wing to the tip of the other. The other cherub also measured ten cubits.*

*Both cherubim had the same measurements and the same form.*

*The height of one cherub was ten cubits and so was the height of the other cherub.*

That specification is one of the earliest known. In addition to its historical interest, something else about it is of note: It has faults. It does not provide any manufacturing tolerances and it is repetitious. It could be said to be a construction industry specification—an industry not noted for high-quality specifications, although the situation is said to be improving. The problem caused by lack of tolerances in this specification is highlighted by the fact that the cubit varied between 18 and 22 inches according to the country and the period. Could that lead to interchangeability problems?

There is at least one other specification of note in the Old Testament, which could be said to belong to the food industry. From Leviticus 11:3:

*Whatever parts the hoof and is cloven footed and chews the cud, among the animals, ye may eat.*

That is a very clear statement of requirements. Unfortunately, however, the clarity of that statement is blurred by the following qualification:

*Nevertheless, amongst those, ye shall not eat these:*

*The camel which parts the hoof and is cloven footed and chews the cud, is unclean to you*

*And the rock badger . . . is unclean to you;*

*and the hare . . . is unclean to you;*

*and the swine . . . is unclean to you.*

This qualification to the original requirement introduces confusion. For example, the hare does not part the hoof. I am not an authority on animals, nor could I get any help from the *Children's Britannica,* but certainly one of those unclean animals does not conform to the stipulated exclusion requirements. The message from this example is, clearly: *Do not qualify otherwise clear statements.* There may also be a translation problem. That is dealt with in later chapters. It is clear that specification problems are not new.

One can find other early examples of processes that can be presumed to have had specifications, although it is more than likely that in the earliest times at least, most of them were oral. Very few people could read or write. One will never know for certain how long it took to develop those earliest processes and to determine the most suitable specifications. But it surely must have taken a long time before those early workers could be sure that their processes were consistent and reliable.

One example is brickmaking. About 1500 B.C., in Egypt, bricks were made by mixing mud and straw and then letting them dry in the sun. As time went on the Egyptians discovered that if they baked their bricks in a kiln, the results were much superior. They were more damp resistant and much harder. But it would have been necessary to determine the maximum and minimum times and temperatures for firing in the kilns. Without this determination the bricks would either be burned or be soft and porous.

Another example from Egypt is that of glassmaking. Originally, it had been found that sand and ash heated together produced glass. Then it was discovered that the ash had to have a high potash content. For many centuries there was little change until, comparatively recently, limestone was used as one of the constituents. In addition, for thousands of years all glass was made by hand, and some still is. It is only within the last hundred or so years that glass has come to be made by machine, with close process control and precise specifications.

Some of Leonardo da Vinci's pigments have been lost to us because of the apparent absence of written specifications. At Hammershus on the Danish island of Bornholm, in the middle of the Baltic Sea, there are 1000 year-old castle ruins. In these ruins the bricks have been eroded by time, but the mortar has not. Today, no one knows what the specification was for this extremely hard wearing mortar. Hydraulic cement was used by the Romans, but the process was lost with the decline of the Roman Empire. It was rediscovered in the mid-eighteenth century when Smeaton built the Eddystone lighthouse. Then, about 1824, Aspdin discovered the process for producing portland cement. It has remained virtually unchanged ever since.

Originally, all these specifications must have been oral and were handed down through the generations. Oral communication of this type is neither safe nor satisfactory. It is not surprising, therefore, that many valuable processes were lost and have not been rediscovered. A game that was popular at family parties during my childhood illustrates very well the problems that can be caused by the oral transfer of information from one person to another, especially when it is between people of different generations. Even with written specifications this can be a problem when, as is so often the case, the users choose to rely on their memories instead of referring to the written text.

In that childhood game the players sat on the floor in a ring. One player would make up a short sentence and whisper it into the ear of the next person. This person passed it on again, and so on, until "a" message came back to the originator. No repetitions were permitted, and each person had to pass on what he or she thought had been said. Invariably, the message that came back to the originator bore little or no resemblance to the original message. The game always provided a lot of fun. Memory applied in a similar fashion to a specification does *not* provide fun. It costs money.

LOST COSTS.

In comparatively recent times, during the reign of Charles I in England, there were written specifications laid down by the Master General of the Ordnance (MGO) for the ammunition supplied for the use of the forces of the king. Some of these specifications have been on view at the old headquarters of the Royal Engineers at the Woolwich Arsenal. (Incidentally, the post of MGO is still in existence, and the Prince of Wales served part of his army career on its staff).

In the early days of industrial history, there was not much direct measurement. For one thing, there was no precise standard of length and no accurate means of measurement. In medieval Germany the foot was equal to the average length of the feet of 12 men emerging, successively, from church on Sunday. The foot was also an ancient measure for black tin, a metallic ore, and equaled 2 gallons. That measure is now nominal and about 60 pounds in weight. The Celtic, or natural foot was equal to 9.9 inches. Three barleycorns equalled 1 thumb; 3 thumbs equalled 1 palm; 3 palms equalled 1 Celtic foot. The yard was equal to 12 palms or ½ fathom or 16 nails. The mile was equal to 1408 ells. It is perhaps no wonder that measurement was neither easy nor accurate.

Much more recently there was developed an international inch, which was derived from the international meter. It is equal to precisely 25.4 millimeters. But there was also a British "standard" inch and an American "standard" inch. All three were different from each other, although not by much—just a few millionths of an inch. This did not cause any problems in ordinary day-to-day measurement but was a cause of difficulty in international work. Resolution of the problem was achieved in 1959. In that year the United Kingdom, the United States, and the countries that used the international inch reached agreement, that the international inch would be used by all countries and would become truly international. Not only in medieval times were measurements difficult to standardize.

In the not too distant past it was common for goods to be sold on the basis of a bargain struck between the seller and the buyer. However, the meaning of words, even then, was subject to considerable change, and the law recognized that fact. This was made crystal clear by the concept of caveat emptor: let the buyer beware. It was up to the buyers to satisfy themselves of the quality of a bargain before it was sealed. That is a maxim that still holds, as many people have found to their cost.

A good present-day example of caveat emptor is that of the street trader whose goods are openly displayed for all to see. A seller may have on display boxes of fruit of excellent quality and good size. A purchaser who buys a quantity of that fruit is liable to have the purchase made up from stock taken from boxes not on display and of lesser quality. It is up to the purchaser to insist on being served from the boxes of high quality fruit being displayed. If the customer does so, the seller cannot, by law, refuse, as the fruit is openly displayed for sale.

Quite early on, it was customary for sellers and buyers to strike a bargain on the basis of agreed samples that would be offered by the seller. The samples would be examined by the buyer and accepted if the buyer was satisfied. This practice is also frequently used today. One well-known industrial example was the British Admiralty pattern. With the use of samples it was easy for the buyer, especially if in possession of an approved copy, to ensure that the goods supplied conformed to the *specification* (the sample) that had been agreed upon. Three very common examples in general use today are cloth, wallpaper, and paint, the first two usually for appearance, pattern, texture, and color, and the last for color only. For none of these does there appear to be a simple alternative, so their use is likely to continue for a long time to come.

In many kinds of industrial products, which are built up from individual parts, interchangeability was a problem. Perhaps it would be more correct to say that interchangeability was *not* a problem, as it was not recognized. There was no guarantee that parts made later would fit an earlier product or even that these parts could be used as replacement parts for a current product. Chronometers are an excellent case in point. A damaged one would be returned to the makers for repair and replacement parts would be specially made to fit.

However, the principle of interchangeability could not come into its own until a means of precision measurement became available. Otherwise, it would not be possible to measure parts with sufficient accuracy to ensure interchangeability. It was probably not until the latter part of the nineteenth century, when the micrometer was developed, that it became possible to make measurements of sufficient accuracy for the purpose. Then interchangeable manufacture became a reality.

Early in the twentieth century Johansson blocks, or slip gauges, were developed. In the United Kingdom in about 1914, the Eden "Millionth" Comparator was developed at the National Physical Laboratory. This instrument was still in use as recently as the 1960s.

Despite these developments there is still selective assembly, even actual fitting. As demand grew, so did growth in the size of the manufacturing organization necessary to deal with it. With growth came the inevitable separation of the various functions within the organization: design, sales, purchasing, engineering, manufacturing, inspection/quality activities. They all became separate and recognizable parts of the whole. All this, of course, helped to speed the increase in the number of specifications in use.

Their use became highly desirable not only for manufacturing purposes but also for buyer protection. Buyers were becoming increasingly less knowledgeable about the details of the products they bought. Reliance on the specifications grew steadily until we find at present that there are many different kinds of specifications used for different purposes. This reliance applies to industrial buyers as well as to private buyers, although in the case of the latter national standards are often very appropriate.

Despite this reliance on specifications, it can be taken that, in general, nobody seems to be very much concerned with the quality of specifications. Although not done primarily to ensure the quality of specifications, Japan took certain steps in this direction over a quarter of a century ago. In Export Inspection Law Number 97 of 1957, steps were taken to ensure that exported product conformed fully to the relevant specifications. A major outcome of this law was a complete reversal of the pre–World War II situation, when Japanese goods were regarded as cheap and shoddy. They are now generally regarded as cheap and good. Even a nonmarket economy such as the USSR is following a similar policy for exported goods.

# 2

# Current Specification Types and Their Purpose

In today's industrial situations specifications are used widely for many purposes. In high-volume and high-technology industries many different types of specifications are in use. The names given to them may well differ from one organization to another but they should be recognizable from the titles and descriptions that follow. It is certain that anyone who is asked to produce a list of the various types in general use would produce a list different from that of anyone else. In an effort to provide some degree of rationalization, the British Standards Institution (BSI) has identified no fewer than 12 different types of specifications which are listed in BS 4778. Even this list is considered to be incomplete, however. I believe that there is at least one additional type that should be included, and readers may well think of others. The 12 types are:

1. Target
2. Functional
3. Product
4. Materials
5. Process
6. Inspection
7. Test
8. Acceptance

9. Installation
10. Use
11. Maintenance
12. Disposal

The additional type that I would suggest including is

13. Procurement

The intended use of each will be discussed, together with comments on the areas of responsibility for their preparation and the interrelationships that exist between types.

**1.   Target specification.**   The purpose of this type is to detail the principal considerations that should be taken into account during the product design stage. Sometimes these are called a design brief, a statement of general requirements that designers should have before they begin work on an actual design stage. A brief should also provide any additional data that may be needed to enable the design of a product that will satisfy all of the stipulated requirements.

There may be differences of opinion about who is responsible for the preparation of this type of specification. The general organization of a company will have an effect, but in most cases it is probably a marketing responsibility. Generally, there will also be an accepted need for engineering participation. Although this type might be expected to stand on its own, thought should be given to the possible dependence on it of many of the other types.

**2.   Functional specification.**   For every product, there must be documentation (which might not be regarded as a specification) that describes the intended functional requirements for a product, and to some extent its limitations as well. For example, a functional requirement for a newly designed electric drill might be that it should drill holes up to ½ inch in diameter in steel and 1½ inches in wood. It is clear that those who prepare many of the other specifications will have to bear this one in mind. A most likely originating source would also be the marketing department.

**3.   Product specification.**   For manufacturing purposes this is the basic specification, as it describes the product to the extent necessary to make it. It will include the details that are required of the various departments involved in the actual manufacture. It is most likely that this specification will be prepared by the production engineering department or its equivalent. It will be based on the target, functional, and materials specifications.

**4. Materials specification.** This is one of the most obvious types as far as purpose is concerned. There may be many of them for a single product, each dealing with a single material that is to be used in the manufacture of the product. The specification should include details of any associated processes that may be used, and will be closely related to the target and functional specifications. Basically, this specification is most likely to be prepared by the design department but with strong involvement on the part of the metallurgy and chemistry laboratories. The engineering department might also be called upon for advice.

**5. Process specification.** According to BS 4778, this specification may be divided into three subgroups, which would be used for discrete items, bulk commodities, and for plant, respectively. Described are the actions that are to be performed during the processing of the raw materials to bring them to the final product stage. They will be closely related to the materials specifications and prepared by the production engineers.

**6. Inspection specification.** Like the material specification, the purpose of this type is fairly obvious. The nature and content will consist of the detail of the various inspections that have to be carried out on the product at various stages of manufacture. Consequential actions that arise from the results of the inspections may also be included. It is likely that for more complex products there will be a series that will apply at successive stages of the manufacturing process. Normally, responsibility for their preparation would rest with the personnel of the inspection or quality departments. In the most advanced industrial organizations responsibility may well rest with personnel in the manufacturing departments. Whoever does the preparation will have to pay close attention to what is detailed in the target and functional specifications, in particular.

**7. Test specification.** Basically similar to the inspection specification, the test specification details any tests that may be required during the manufacturing and inspection stages. It will, of course, also deal with any consequential actions that may arise from the results of the tests. As in the case of the inspection specifications it is likely that there will be a series of them for more complex products. Responsibility for preparation will generally be as for the inspection specifications as will relationships with other types.

**8. Acceptance specification.** This type sets out the various criteria that will govern the acceptance of a product at various stages of manufacture and at final acceptance upon completion of manufacture. According to the complexity of the product, the size may range from a single page to many pages. One such example is described in section 2.7 of chapter 5. Their preparation should be the responsibility of the inspection/quality department, although, again, when complex products are concerned, other departments may be involved, for example, the service and engineering departments. This may depend on whether the specification is intended for an internally produced item or one that is bought outside. In any event, this specification is likely to be closely related to target, functional, material, inspection and test specifications. It may also have links with process specifications.

**9. Installation specification.** Although basically a specification, this type is more likely to be recognized as an installation manual. Of all the types described so far, this is the first that will not be required for simple products. Only when products of an increasingly complex nature are concerned will this specification be required. All instructions necessary for installation of the product on site ready for operation will be included. Site preparation may also be required prior to installation. The preparation of an installation specification will almost certainly be an engineering responsibility with some design involvement. This specification will have relationships with target, functional, and product specifications. There will also be a close relationship with the next type described, the use specification.

**10. Use specification.** As in the case of the installation specification, this one is likely to be recognized more readily as a user manual. It will contain all the information, including any special instructions, that users will need to enable them to set up the product for use in the intended manner. For many products there will be no need for this type of specification, and for others its provision will be optional. But there will be many products for which it will be essential, a very relevant current example being the home computer. The marketing department will probably be responsible for preparation of the specification with design and engineering involvement. This specification will be related to the target, functional, product, and installation specifications. It will also have a relationship with the next two types described, the maintenance and disposal types. In many cases it is likely that types 9 to 12 will be found together under the umbrella of a user/operator manual.

**11. Maintenance specification.** This type describes the detailed procedures that should be followed to ensure that a product receives correct maintenance at the required intervals. Responsibility for its preparation will probably lie with the engineering department, with some design and possibly, marketing involvement. Attention will have to be paid to one or more of the target, functional, product, materials, installation, and use specifications.

**12. Disposal specification.** This type is relatively new to industry, although it has been familiar in the armaments field for a long time. More and more products are becoming available for which specialist advice, and sometimes facilities, are required when disposal is necessary. These products are too dangerous simply to be thrown onto a rubbish pile, although recognition of that fact is comparatively recent in many cases. Particularly obvious examples are poisonous chemicals and radioactive materials. Much commoner and within the experience of the population at large are aerosol cans and fluorescent light tubes. There is another modern product which, although little known outside its own field, requires specialist disposal instructions: the prestressed concrete structure. When such structures come to be demolished, nobody is quite sure what will happen, except that considerable stress will be released suddenly. Nor is anyone quite sure how to go about the demolition.

Whatever the product, these specifications describe the steps that must be taken when it is necessary to dispose of the product, either because it has reached the end of its useful life or because it has failed prematurely. The steps will range from simple to complex and dangerous. Responsibility for their preparation will probably be shared between design and engineering. The main relationships are likely to be with the material specifications, although some regard may also have to be given to the installation, use, and maintenance specifications as well.

Those are the 12 types listed in BS 4778. The additional type is now discussed.

**13.   Procurement specification.**   This is a type that is common in some industries, of which electronics is a particular case. It is usually looked upon as a kind of general specification which includes within it elements of some or all of the first eight types identified in BS 4778. An example is that of a buying organization that wishes to procure a relatively complex product which has to meet a wide variety of requirements. These may be in the materials, functional, inspection, test, and acceptance areas. They provide a total picture of the what, where, when, and how expected of the item so that prospective suppliers know what is expected of them. Procurement specifications are usually a joint venture of several departments, including quality, design, and engineering, if not others.

**Additional types of specifications.**   Although the fact is not generally recognized, it can be said that *any* instruction is a form of specification. This must be so because any instruction is the specification of an action that is to be carried out. For example, a production manager requests a staff member to make an investigation into the causes of an unexpected rise in the percentage of process scrap. This request is a specification for a particular action. The specification requires the assistant to go to the manufacturing department to carry out an investigation and come back with the results. The manager will discuss the results with the assistant to be sure that the instructions have been carried out properly.

Except that they are obviously more complex than a simple instruction, an operational procedure is also a form of specification, and industry relies very much upon them for its operational efficiency. Most companies have operational procedure manuals. They describe sets of requirements, which is what a specification does. They usually set out criteria by means of which conformance to the requirement can be verified—which is what a specification does. They may describe corrective actions to be taken in appropriate circumstances—which is what a specification may also do.

Standards, whether company, national, or international, are another form of specification, a type known to most people. Perhaps the best known in the United Kingdom are the British Standards, published by the British Standards Institution, of which there are nearly 10,000. Most developed countries have their own, the U.S. American National Standards Institute, the West German DIN (Deutsche Institut für Normung) and those of the International Standards Organization being among the best known. Readers will be able to identify other types of specification standards which are known to them.

There is a need for a wider recognition of the fact that the *faults* that appear in specifications, as that word is generally understood, are exactly paralleled in other forms of specification, as are the ways of avoiding the faults. The general consequences that stem from faults in specifications, the lack of quality in a product, will also apply to those other forms, except that the product, and the measure of its quality, are rather different.

# 3

# Basic Specification Problems

Specifications are a fruitful source of problems. During the American *Apollo* space mission, which was reputed to have incorporated some 15 million components, a number of technical problems arose. A number of later missions had similar problems, and in several cases the consequences were disastrous. A nuclear submarine was lost in the Atlantic Ocean for similar reasons. "Technical problems" have held up the programs for both the American and Russian space shuttles and the European Arianne rocket. It would be foolhardy to believe that specifications did not have a part in those problems. In fact, one can say with a fair degree of confidence that many of these problems could be traced, in one way or another, to poor quality specifications. Just before he blasted off into space on the first manned U.S. mission, Colonel Glenn thought, as he said later: "Here I am sitting on the top of thousands of critical components and all of them are made by the lowest bidder!" (Did the bidder always understand the specifications?)

During the last three decades there has been a significant increase in the technical complexity of specifications. This relates not only to the increase in the technical level of the requirements which they state but, also, in the great length of many of them. Perhaps as an almost automatic consequence it is also apparent that users of these specifications find it increasingly difficult to understand them as they are intended to be understood. Nor can the users always correctly gauge the intention of the writer. If this intention is not properly understood, then, indeed the specification is one of poor quality.

"This product is of the highest standards of quality." This is a statement which, in one form or another, is quite often made in product advertising. Apart from the fact that this phrase does not mean very much as it stands, it is most surprising to find that industry gives little, if any, thought to the quality of the specifications on which it relies. Yet it depends on them for the production of those very products of the highest standards of quality. *Poor-quality specifications must inevitably result in poor-quality product.*

It is possible that one reason for this situation is the declining standards of basic language teaching in modern schools. There have been several recent official reports deploring this fact. Perhaps even more surprising is the statement attributed to one teacher in response to a question from an anxious parent: "We don't bother about grammar and spelling nowadays."

Quite often in factories it will be found that processes and their controlling specifications do not quite match. Over many years I have seldom found an exact match between the two when carrying out an audit. The deviations range from minor to major, but even the minor ones are undesirable. Customer specifications in particular are not always fully understood by the supplier. As a result, one of the following two situations is likely to arise:

1. The detail requirements will have been ignored by the supplier, with the eventual result that costly "quality" problems arise. (All problems cost money.)
2. The supplier will carefully scrutinize all incoming customer specifications. If he or she see things that he or she does not understand or like, he or she may refer the doubt to the customer. Such a reference will, in all probability, result in an expensive series of meetings. Changes may be made in the specifications, thus producing the improvement that should have been there in the first instance.

Alternative 2 is the one in which costs *obviously* arise. However, the alternative most likely to be adopted is the first. Both situations are serious. What is even more serious is the fact that the additional costs which inevitably arise are not usually suspected—at least, not as having been caused by a specification problem. These additional costs will be greatest in the newer, high-technology industries if only because of the sheer volume of the number of specifications that are prepared and have to be used. It must also be remembered that many high-technology companies are small ones that will, more often than not, be using specifications prepared by the larger organizations. So small companies as well as large ones are affected.

It should not take much thought to realize that one reason for the apparent neglect of such an important factor is lack of knowledge. Nowhere in the accounts of a company will one find information about the costs that arise from the preparation of poor quality specifications. At least, they are never shown as such. Of course, there will be a cost for a section, perhaps of the engineering department, that has an official responsibility for the preparation of specifications. It might, for example, be called the *standards group*. Its costs will be known with accuracy. Or will they?

Nowhere in the accounts will there be a cost for the revision of specifications. Nowhere in the accounts will there be a cost for the correction of errors caused by the use of specifications of poor quality. Rather, there will be costs under quite different headings, such as scrap, rework, and rectification of product. These costs will usually be held to be the responsibility of the production department, *not* of the specification writers at whose door they should lie.

Lost Costs.

In recent years there has been some discussion of the subject, both in the pages of a few technical journals and at international conferences. In these discussions some remedies have been proposed but, unfortunately, little has actually been done to correct the situation. More than likely this is because, as the costs incurred—the LOST COSTS—are concealed, it is difficult, if not impossible, to convince management of their existence.

To simplify consideration of the problem in this book, a considerable number of types of faults in specifications have been identified over the years. They have been divided into three groups of related faults, but no particular level of importance should be assumed from the position of any fault in its group. This list is not all-inclusive, nor will the grouping be agreed to by everyone. The groupings are only for convenience. Readers may feel that some of the examples might apply equally to a different fault group than the one indicated.

The groups appear under the following headings;

*Group 1*—Words
*Group 2*—Ideas
*Group 3*—Presentation

These headings should be sufficiently descriptive for positive identification. They are dealt with in Chapters 4, 5, and 6, respectively.
In the discussions there are many examples of the faults, together
with commentaries.

Group 1—Words:

1.1 The use of words in nonstandard meanings
1.2 Changing meanings of technical words
1.3 Inconsistency in the use of words and phrases
1.4 Grammar
1.5 Incorrect text references
1.6 Definitions
1.7 Ambiguity

Group 2—Ideas:

2.1  Lack of clarity/confused statements
2.2  Repetition
2.3  Contradiction
2.4  Nonsense statements
2.5  "Dangerous" extracts from other documents
2.6  Excessive requirements
2.7  Multiple mixed specification systems
2.8  Chain specification systems
2.9  Multiple authors
2.10 Omissions

Group 3—Presentation:

3.1 Lack of consideration for the user
3.2 Unbelievable examples
3.3 Untrue statements
3.4 Philosophizing
3.5 "Reading across" and illogical ordering of contents in a
    specification system
3.6 Poor visual material
3.7 Unsatisfactory proofreading
3.8 Lack of an index in larger specifications

At this point it is worth looking a little more closely at the idea of the quality of a specification. There are many, long-established, successful methods which are used to determine the quality of a product. But there is no equivalent method that can be used to establish the quality of a specification. In fact, seldom, if ever, is the quality of a specification considered. If a method is ever developed, it will likely be one that is highly subjective and open to question. In the meantime it is suggested that there is one possibility, although it is one that can only be applied after the event, so to speak.

If a specification has been in use for some time and has not caused any problems in use or understanding, it is probably of a good standard of quality. Another important point that should be borne in mind when preparing draft specifications is that they are often given a preliminary circulation to persons both inside and outside the organizations that prepared them. These people are invited to comment on them from the standpoint of their knowledge and experience. For this to be an effective scrutiny, the draft must be read carefully and fully understood. This becomes almost impossible if the draft is of a style and the presentation of such quality the readers are unable to carry out their task and give up in disgust. This deprives the issuing organization of valuable and informed comment and makes its job much more difficult.

It is seldom recognized that a major task for a specification-issuing organization is to produce specifications that are of a satisfactory standard of quality. This relates not only to a clear statement of the requirements but also to the ease with which those requirements are understood by those for whom the specification is intended. Note that these two points are not necessarily the same. For example, a requirement may be set out very clearly indeed in a high-level language easily understood by a Ph.D. in the same discipline. However, if the main users of a specification are to be shop floor personnel, it will almost certainly *not* be understood. Practically speaking, every industry suffers from this fault at some time or other.

## The EEC Directive Relating to Product Liability

The directive of the European Economic Community (EEC) regarding product liability was first published in draft form in September 1976, with a revised version in September 1979. The final version was issued at the end of July, 1985 as a full directive of considerable importance to that large part of the manufacturing industry that produces consumer products. It is discussed here briefly because it has much the same effect as a specification and because it clearly demonstrates many of the faults that are to be found in specifications, particularly in the choice and use of words and the problems that arise when a number of languages are concerned (as happens with companies operating in several countries).

The wide range of languages within the EEC, the problems of translation and the ambiguity of many of the technical/legal words and phrases used all raise many problems. Some of them will be familiar to specification writers whose companies deal with multiple language specifications and relate especially to the choice and use of words. It is not uncommon in specifications presented in more than one language to see a statement such as the following: "If any question arises concerning the interpretation of a requirement in this specification, the definitive version is the English language version." Such a statement cannot be applied to EEC documents and directives, because in the EEC, all languages are equal. This means that extreme care must be taken when choosing words.

It is believed that in the case of the draft directive under discussion, the working documents were prepared in German. All the other language versions were then translated from the original German. However, the problem is not just the actual translation but also relates to the selection of suitable equivalent words. I do not know if this is a problem in other languages, but in English a very wide choice of words is usually available, which permits a considerable range of shades of meaning.

A very good example is to be found in Article 6, which sets out the proposed definition for a defective product. It has commonly been understood by industry for many years that a defective product is one that does not conform to its governing specification in a measurable way. (This is more or less in agreement with the definition in BS 4778.) The degree of nonconformance can be expressed precisely in numbers relating to particular parameters. The EEC commission has turned this long-standing belief on its head. The proposed new definition is as follows:

*A product is defective when it does not provide for persons or property the safety which a person is entitled to expect.*

It does not take much thought to appreciate the many meanings that can be read into this definition, and the lack of precision that it displays, especially with regard to the vagueness in meaning conveyed by the word "persons." In the 1979 revised draft of the directive, a similar complication arose from a small change made by the commission to "help" manufacturers. In the original version the likelihood of an injured person being able to claim compensation if he or she had used a product for a purpose other than that intended by the manufacturer was greatly reduced. However, the commission added the word "apparent" before the word "use" and explained their thinking in the following manner:

*Apparent use is intended to relate to the use for which users of the product consider it most useful, rather than the use that the manufacturer had intended.*

Perhaps the best example of apparent use as opposed to the use intended by the manufacturer is that of a screwdriver. Almost everywhere a very common use for a screwdriver is not to drive a screw, but to open a can of paint. In California, some years ago, this resulted in a substantial award of compensation to a claimant. He had been using a screwdriver to open a can of paint when the screwdriver blade shattered. A splinter of steel entered his eye and caused serious injury. The blade had apparently been overhardened. For its intended use this would not have been much of a problem as the operational load would have been one of torque. But in its apparent use, the blade would not take the considerable bending load caused by trying to open the can of paint.

A further use of an inappropriate word is found in Article 16 which reads as follows:

*... may provide that a producer's total liability for any damage resulting from a death or personal injury and caused by identical articles with the same defect shall be limited to an amount which may not be less than 70 million European Units of Account.*

(an amount of approximately $75,000,000, depending on the rate of exchange).

The word "identical" is a strong word in the English language. Also it is well understood in manufacturing that there is no such thing as an identical article. Normal manufacturing variations account for that. However, if one tries to be reasonable in interpreting the meaning of the word (which should never be done with a legal document), additional problems arise. This is because manufacturers would have to insure against possible liability for compensation that might arise from such "identical" articles. Suppose that a manufacturer produces a range of articles that are basically similar but different in appearance and size—perhaps a range of bicycles. The manufacturer may list in its catalog hundreds of different models which are not identical. Does the manufacturer have to insure for the limit of $75,000,000 for each of hundreds of models? The bicycle models may be similar, but are they identical?

In the Dutch and German language versions of the draft directive, the words used in the place of "identical" mean "similar" and "like", respectively, when translated back into English. From the point of view of both manufacturing and insuring, the use of either of these two words in the English-language version would simplify the problem considerably. Then the manufacturer would only have to insure for a single sum of $75,000,000. It may be of some interest to know that the choice of the word "identical" is said to have been made by a professor of English at Cambridge University and a senior civil servant. One might hazard a guess that two more unlikely persons to understand the significance to industry of the word "identical" for such a purpose would be hard to find.

There is another awkward consequence that could arise from the intent and the wording of Article 16 in the draft directive. It is an excellent example of fault group 2.6 in Chapter 5: excessive requirements. If a range of identical articles has been placed on the market and is subsequently found to possess a single type of defect that causes harm to users—perhaps a range of overhardened screwdrivers—there is likely to be a series of claims for compensation. The total amount that can be paid in these circumstances, according to the directive, is $75,000,000. So as the claims come in there are two alternative courses open to the insurers:

1. As each claim comes in, it will be paid until, one fine day, it will be necessary to say to a claimant: "Sorry, there is no more money available to pay you. The law says that there is a

limit to the amount of compensation that can be paid, and we have reached that limit".

2. As each claim comes in, the claimant will be told that he or she will have to wait until it is clear that there will be no more claims (who would know how long that might be?) before any decision could be made as to the amount that could be paid.

Of course, both alternatives are ridiculous. An answer will eventually be found in the courts, but this is an expensive method of problem solving. It would have been so much better if a little more thought had originally been devoted to the possible consequences of such a stipulation.

There are a number of similar problems in this directive relating to the choice, and use, of words. It is, of course, a form of specification—even if a rather special one—which is why it has been included here.

*Note:* On July 30, 1985, the final directive was issued. A number of changes were made, of which two major ones resulted in the removal of the "apparent use" requirement, and optional inclusion of the "identical" clause according to the wish of each member state. (The United Kingdom has excluded it.)

# 4
# Specification Faults: Group 1—Words

I know that you think that you understand what I said. But what you do not understand is that what I said is not what I meant.

## 1.1 The Use of Words in Nonstandard Meanings

If one is writing a novel of any kind, there are advantages to be gained in word substitution as the story progresses. It adds to the interest and takes advantage of the very great word choice that is available in the English language. After all, variety is the spice of life. Sometimes, too, it is permissible to use a word in an unusual way for special emphasis. All these things can be done in a story without any harm being done, except, perhaps, to the feelings of a philology purist. After all, if the reader of a novel sees a different meaning in a word from that which was intended by the writer, no great harm has been done.

When one is writing a specification, or a standard, an operating procedure, or an instruction, however, it is of the utmost importance that words be used in the normal, or generally understood, sense. It is not just that the readers or users of a specification have to understand what a word means to them. They have to understand what the writer wants them to understand. It is therefore essential that they are able to read into a word exactly what the writer wrote into that word. Consider the following examples.

### a.  Relating to an inspection instruction

*If a lot is withdrawn in a state of failure...*

One is tempted to believe in this case that the writer was trying to create a new kind of failure. Perhaps he was, but in the general context in which the statement was made, it is clear that a simpler and more readily understandable statement would have been:

If a failed lot is withdrawn...

### b.  Relating to environmental conditions

Fungus-resistant material. *External parts of the [component] shall be inherently nonnutrient.*

In this case it would have been better and simpler to have said, perhaps:

... shall not promote the growth of fungus.

### c.  Relating to marking requirements

*An index point shall be so marked that it can be read from either the top or the bottom of any component package in accordance with standard marking procedures.*

From a later paragraph:

*Components which have passed the detailed inspection of this specification shall be marked with an emerald green dot so that it will not be confused with the ... orientation dot.*

Although it is not easy to provide an index point that can be seen from the top or the bottom of the component package, it will be rather more difficult, to say the least, to mark an index point to meet the same requirements.

The second paragraph makes it clear that the index point is regarded as a paint spot, although the "index point" had changed to an "orientation dot." Another interesting point is that the second paragraph is taken from an appendix to the specification which dealt with radiological inspection only. The form of wording suggests that the emerald green dot indicates compliance with the requirements of the entire specification (which it does not) instead of the radiological requirements of the appendix. Also, the words "so that it" before "will not be confused" could with advantage be changed to "which." It will not be surprising for readers to be told that this specification was written by three persons who did not collaborate too well. This example could have been used in section 2.9 in Chapter 5.

### d. Relating to a definition of failure

*Failures are escaping bubbles caused by FC 78 expansion within the package.*

In this case the symptom seems to be regarded as the fault. It would be much better if this definition were changed as follows:

Failures are indicated by escaping bubbles ...

### e. Relating to records

*Any such material lost, damaged, or otherwise unsuitable for use shall be recorded and reported to the customer.*

This requirement related to free-issue material; a simpler way to express the requirement might be:

*A record shall be kept of any such material which is lost, damaged, or is unsuitable for use and a report made to the customer.*

### f. Relating to the calibration of instruments

*No matter how invariant an instrument may be ...*

Apart from the fact that use of the word "invariant" is most unusual in this context, it would be easier for most people to understand if the word "stable" were used in its place.

**g. Relating to definitions.** In an explanation of the concept of failure classification, there is the following statement, which discusses the identification of the various types of failures:

*This is done in this document by adding an adjectival modifier to the word "failure".*

Later examples of the adjectival modifier are "misuse failure," "inherent weakness failure," and "wearout failure," to mention only a few. All that it should have been necessary to say was, for example, in the definition:

*An adjective is used before the word "failure."*

### h.  Relating to philosophy

> *... the simplest way of effecting a decision ...*

This really means:

> *... the simplest way of reaching a decision ...*

## 1.2   Changing the Meanings of Technical Words

These faults are closely related to those of section 1.1 but are concerned more directly with the use of technical words. In addition to specifications, this practice is to be found in many other types of documents: for example, in scientific reports. A clear example of this kind is "depressed socioeconomic area" for "slum."

### a.  Relating to electrical measurements

> Endpoints. *Endpoint measurements shall be recorded before starting ...*

One would have thought that endpoint measurements would come last. Attention is also drawn to what is a very common fault—repetition. The first words of a paragraph, used as a form of title, are immediately repeated in the following text: in this case the word "endpoint."
Later in the same specification:

> *Electrical parameters used as endpoint measurements ...*

There is much that is wrong with this example. A much better form of words would be something like this:

> Specified values of electrical parameters will be used for endpoint measurement limits.

### b.   Relating to work instructions.  Under the heading of "Work Instructions," the following appeared:

> *Clear and precise documented instructions ...*

There was a later heading "Inspection Procedures." Although not all work instructions are inspection procedures, all inspection procedures are work instructions. In other parts of the same specification there appeared a number of other forms of words that could be construed as meaning "inspection procedures."

### c. Relating to basic principles

> It should be realized, however, that only a statement of management procedures is possible ...

Use of the word "procedure" in this sense is quite incorrect. The appropriate word is "principles." Although a statement of management principles is possible, it would be very difficult to make a statement of management procedures. One might, however, make a list of management procedures.

### d. Relating to a degree of confusion

> ... during the normal running of the process ...
> ... and a normal distribution ...

There is a difference in these special cases in the meanings of the words "normal" and "Normal." In the first of the two examples the word "normal is used correctly. In the second example it should be "Normal distribution." That is the form used in statistical work to distinguish distributions that are Normal from those that are *not* Normal.

## 1.3 Inconsistency in the Use of Words and Phrases

This is a particularly irritating fault because it seems to suggest a general sloppiness on the part of the writer of a specification. It could also mean that there are a number of other faults and this is often the case. One example that has appeared in a number of American specifications is that of "cognizant engineer." A more appropriate term would be "relevant engineer." However, of the many examples which I have seen, the first one following is, arguably, the best of its kind.

**a. Relating to basic principles.** In a complex specification there was a requirement that could have been expressed in the following words:

> ... will establish and maintain procedures that will clearly describe the method to be used for ...

Many other phrases could have been formulated to describe this requirement. However, in this particular specification no fewer than 12 different forms of words were *intended to* describe the requirement. They were as follows:

1. Establish and maintain procedures
2. Establish and maintain a system
3. Establish and maintain control
4. Be responsible for
5. Maintain control of
6. Ensure the ability
7. Arrange for
8. Provide
9. Establish a procedure
10. Develop and maintain
11. Ensure that
12. Establish a system

These 12 phrases are given in the order in which they actually appeared. They certainly represent a variety of ways of trying to say the same thing in different places in a specification. The result is that the real requirement is blurred because of the many variations on a theme.

The same specification also contained the following phrase:

*Procedures shall ensure that ...*

Of course, procedures in themselves cannot *do* anything. They should tell the reader how to do it. Also the word "shall" although a command word, means a promise or a determination. According to the *Oxford English Dictionary*, the word "will" means an intention or determination that another or others shall do something. Many specifications use the word "shall" as a command word. As in every case they are using it in the sense that there is an intention or determination that another or others shall do whatever it is that they are specifying, it would seem that "will" would be the more correct word to use for the purpose.

## b. Relating to the establishment of a design procedure

*The establishment of design review procedures shall ensure progress towards the achievement of the design and development program objectives through the timely identification of problem areas, and the use of defect rate data feedback from previous designs when appropriate.*

First it may be said that this is a rather long sentence, although that is of comparatively minor importance. However, *the establish-*

*ment of design review procedures* will not itself ensure progress. To ensure that progress, the procedures must be used, and this fact is often overlooked. It would have been better and simpler to have said, for example:

Design review procedures will be established and their use will ensure progress toward . . .

## c.  Relating to basic principles

(i)   *Jigs, fixtures, templates, patterns, or other such devices used as a media of inspection* . . .

Apart from the rather odd use of the word "medium" in such a context, it would have been simpler and much clearer to most readers if the word "means" or perhaps "method" had been used in place of "medium."

(ii)  *The responsibility shall ensure that controls required are effective* . . .

As responsibility is an attribute that is possessed by a person or persons, it is a little difficult to imagine how this attribute, on its own, "shall ensure that . . . " The immediately preceding sentence in the same paragraph is marginally better:

*Responsibility shall be established for ensuring that* . . .

In this case it would have been better to say, perhaps:

*The . . . manager will ultimately be responsible for* . . .

## d.  Relating to inspection instructions.  In a section of a specification discussing the subject of decisions in sampling inspection plans, a total of eight types of "decisions" were listed:

Acceptance*
Acceptance number
Rejection/reject*
Rejection number
Fraction defective
Quality level
Limiting quality
Lot tolerance percent defective

As is apparent, only the two items indicated by an asterisk are decisions. The other items could more correctly be described as *factors* that are considered when making decisions in sampling inspection tasks. There is also at least one more decision that should have been included in that listing, but is not. It is

Determine the appropriate sampling plan.

## 1.4 Grammar

It is not likely that the use of bad grammar and sentence construction will cause serious problems in the understanding of specifications. However, as a matter of course, and to ensure that a good text is provided, it should be discouraged. It is, among other things, a pointer toward the general deterioration in writing standards that is taking place. More important, as it is exact understanding that must be achieved in a specification, the use of bad grammar is undesirable. The following are examples of bad grammar and sentence construction that cast a cloud on understanding:

a. *Measurements carried out by the factory, and also measurements carried out by measuring instruments calibrated by the factory, can have their ancestry traced back via a chain of calibrations to a master definitive standard.*

b. *The system should include a necessary measuring equipment privately owned or hired from commercial facilities.*

c. *To carry out a review adequately, it is accepted that procedures should be proposed which will define the conduct of the review, who will perform the review, what will be examined, its periodicity, how and where the review will be done, to how the results of the review will be reported and how necessary corrective action will be instituted.*

d. *Valid estimates of the amount of the error can be estimated.*

e. *The total error, which may be expressed as a percentage, fraction, etc., may be comprised of . . .*

f. *There is an overall requirement for records to be maintained of any errors that exceed the designated limits.*

g. *Similarly, when the overall device is safeguarded from tampering, it is not intended that all components that are subject to the adjustment are individually protected.*

h. *. . . where one or other type of chart is preferable to the other.*

i. *The improvement of the accuracy of the standards ... for measuring equipment used by shortening the calibration chain.*

j. Apparatus. *Suitable electrical measurement equipment necessary to determine compliance with the requirements of the applicable procurement document and other apparatus as referenced in the relevant test schedules.*

Whether the writer of this last gem was called away before completing the sentence and then forgot that fact is open to question. It is more than likely that he got lost in his own words before reaching the end. At the very least, three more words are required to complete the sentence, for example, " ... will be provided."

k. *Once a lot has passed the 1000-hour life test\* tests with minimums of 240 hours and maximums of 2000 hours may be initiated for lots\* provided that 120 days have not elapsed since a 1000-hour test.*

In this case the addition of commas where indicated by the two asterisks would assist in an understanding of what is intended.

## 1.5   Incorrect Text References

Some time ago I came across a technical book with an errata slip which said something like the following;

*In the index please add one page number to page number references after page 145.*

Obviously, someone had dropped a rather large brick here. However, this error is not as uncommon as one might think. In the case of one draft specification that I received for review, it was necessary to make the following comment:

*With the exception of the second paragraph on page 20, every reference to a page number on which is to be found a table, figure, or an instruction note is incorrect; that is, the wrong page number is given. In one or two instances the clause numbers to which one is referred are also incorrect. In several places one is referred to a Clause 4. While there is a Clause 4 in the document, its relevance to the various parts of the text that called it up was not apparent.*

In another place there was a reference to the "top two charts" and "top two diagrams" without any indication of where they were to be found. Later on in the same paragraph there was a reference to a "Figure 6". From the general sense of the text one could reasonably assume that this was the diagram with the top two. But to have a top of anything it is reasonable to suppose that there is also a bottom. Unfortunately, Figure 6 contained only one diagram.

*Another specification contained the following statement:*

Conditioning shall be as required by paragraphs 1.5.1.1 through 4.5.1.8 as required by Table 1. The test shall be conducted on 100% of the lot prior to submission of the lot to the tests specified in Table III.

No location was given for the tables; indeed, there were no tables in the specification. An intelligent guess in this case was that the tables were to be found in a subsidiary specification to the one from which the extract was taken.

Experience suggests that the reason for faults of this kind is careless checking before issue. It is very often the case that a draft specification has been revised before being finalized. There have been additions; there have been deletions; some sections have been moved to other places. No one checks to see that the text references are still correct. When several persons are involved with the same document, it is easy for each to assume that one of the others has carried out the check. It is so easy to miss things like this that extra care should be taken to ensure that they will not be missed.

## 1.6   Definitions

This class of fault is used to deal with various aspects of definitions. Which reader would immediately, or ever, understand that "a task recognizer and problem solver" is actually a designer?

What does one really understand by the term "undetected failure time"? The first response to this question might be: "It must be the elapsed time to a failure that is undetected." Reasonable? But if the failure is not detected, how does one know that there was a failure and therefore that there was an elapsed time? If neither of these facts is known, or can be known, can there be any point in the definition? Not unless this never-to-be-known fact is to be discussed as a purely philosophical exercise. And that is not what specifications are for. As a matter of fact, the definition given for the undetected failure time was as follows;

*The time between the instant of failure and the actual detection of the failure.*

In another example it was stated that explanations for a number of terms would be found in another document. Unfortunately, that document had not yet been prepared. In another part of this specification, there was a reference to 10% and 50% gauges. What does this mean? In a lifetime of experience in the fields of inspection and quality control, I have never heard of them, but they could exist.

Another example:

*(c)* Acceptance number. *(c) The acceptance number is defined as an integral number associated with the selected sample size which determines the maximum number of defectives permitted for that sample size.*

Those who are familiar with the mechanics of sampling inspection will appreciate that there are at least two more important factors that directly influence the acceptance number for "that sample size": the AQL/LTPD and the inspection level. In any case the acceptance number does not determine the maximum number of defectives permitted; it *is* the maximum number permitted.

Two more examples are given without comment. The first is from a process specification that was in use until at least 1980:

*Barrelling of 5½-inch forgings.*
*4 days.    Ordinary corundum.*
*           1 bucket of water.*
*           1 tin sand pebbles.*
*           half bucket of SC 9 oil.*
*           1 cup Brightokleen.*
*Wash off after 4 days.*
*4 days.    Barrel polish in;*
*           1 tin sand pebbles.*
*           half cup SC 9 oil.*
*           1 cup Brightokleen.*
*Wash off after 8 days.*

The second example is taken from a list of definitions relating to the type of reflectors used for bicycles:

*Reference center. The orthogonal projection of the center of gravity of the effective reflex surface on the plane nearest to the observer which is tangential to that surface and perpendicular to the reference axis of the reflex reflecting device.*

## 1.7 Ambiguity

Ambiguity is a fault found in all sorts of places: in newspapers, on radio and television, and even in legislative acts. When present in specifications, it is one of the more irritating faults to be found because of the uncertainty that it raises in the minds of those who have to read and understand the specifications. For example;

*(a)  ... records of inspection operations shall cover ... acceptance, rejection or other final disposition of the material.*

Ordinarily, one either accepts or rejects material and one is tempted to wonder what the third possibility, "other final disposition," might be? It is true that there might be possibilities for rework or rectification, but neither of these is a final disposition. It may be that the possibility was being considered that material not acceptable for military purposes could be used for ordinary commercial purposes. That has happened. However, one feels that if this were the case, that fact should have been noted in the specification. It does not automatically follow that material unsuitable for military purposes will be acceptable for commercial use.

*(b)  ... documentation considered to be proprietary by the manufacturer shall not ordinarily be accessible to the procuring activity; its existence shall be established by a current listing by title, document number, revision number and effective date of the last revision, and its implementation shall be certified by a responsible official of ... upon request by the procuring activity.*

Apart from being a rather long and wordy sentence, this seems to be fairly reasonable in its intentions. The manufacturer is restricting access to certain documents. However, the next sentence reads as follows:

*All required program documentation shall be available by the qualifying or procuring activity upon request, and the qualifying*

*qualifying or procuring activity upon request, and the qualifying or procuring activity upon request, and the qualifying or procuring activity shall have access to nonproprietary areas of the manufacturer's plant for the purpose of verifying its implementation.*

The specification from which these two extracts have been taken was based on a military specification. In this context it will be obvious that the military authorities are always the customer—never manufacturers. There should thus be a very clear distinction between the procuring activity, which is also the qualification authority, and the manufacturer. Unfortunately, the writers of the specification did not seem to have appreciated the fact. Or perhaps they failed to take it into account in an effort to save specification writing time; time saved by extracting large parts of the military specification for their own specification, and, without carefully checking the relevance of the extracts.

Apart from the obvious contradiction, there is also a fair degree of additional confusion. No doubt it is because of this confusion, in which the writers have lost their way, that they seem to be stating some rather curious things!

That they could not have access to certain parts of their own plant.

That they could not see some of their own documentation.

That their own quality staff must satisfy themselves that their own test facilities are satisfactory to carry out the testing which they themselves have specified to suit their own facilities.

Although this does appear to be an extraordinary example of ambiguity and confusion, it shows what can happen. In a related specification in the same series, prepared by the same manufacturer, it was made perfectly clear that the same manufacturer is also, in this particular case, the procuring activity.

Another conclusion that might be drawn from this example is: "Do not sign that which you have not read or do not understand." This particular specification had been signed by no fewer than six people of increasing rank up to vice-presidential level. This specification could also be used as an example in section 2.9 "Multiple Authors," in Chapter 5. It is sad to think of the effort that went into the creation of such a complicated and confusing set of requirements—especially when a more-than-equivalent amount of effort and time must be expended later in an effort to understand it. I never was able to get explanations for these last examples.

MORE LOST COSTS.

# 5

# Specification Faults: Group 2—Ideas

In this chapter the faults discussed could be classed as *idea faults* as opposed to those relating specifically to words. The groupings may seem to be a rather frightening list of "fault" types found in specifications. However, be assured that all of them have been seen and identified and that there are examples of all.

He who writes errors into specifications is really quite generous in the mass. (Old Chinese proverb)

## 2.1 Lack of Clarity/Confused Statements

a. Following is an extract from a complex specification for electronic components:

*The equipment shall be identified in which a different circuit can be used and which is available from a different source which shall comply with any one of the sources given in clauses 1.7.1.1, 1.7.1.2 or 1.7.1.3.*

(None of these clauses gave any indication of the meaning of this statement.) When the originating organization was asked to explain the meaning of this obscure statement, the answer, given with apologies it should be noted, was as follows:

*If equipment is supplied to the [organization] which contains*

**39**

*foreign-made parts the [organization] must be told and will then clearly mark such equipment to show that it contains foreign-made parts. The equipment maker will also be required to assure the [organization] that, if necessary, the equipment could be modified to use alternative, British-made components.*

One feels that the escape clause seen at the beginning of films and some books could have been applied to this example (slightly rephrased):

*Any resemblance between the real meaning and this statement is not even coincidental.*

b. A confusing lack of accuracy is illustrated by this example:

*All penetramater wires, paragraph 7.1, shall be visible in each radiograph.*

This is clear except for the minor fact that paragraph 7.1 dealt with acceptance inspection. Penetramaters were dealt with in paragraph 8. This example could clearly have been used in section 1.5, "Incorrect Text References," in Chapter 4.

c. *Suggestions* for courses of action should not be given in specifications. Stipulated actions or requirements should be given. For example:

*Suggest using Markem Ink No. 7224. . . .*

Instead of this permissive statement, the writer should have said, perhaps:

*Use Markem Ink No. 7224. . . .*

d. Say what is meant. One should always be sure that requirements not only mean what they say but say what they mean. (Consider the reminder that appeared at the beginning of Chapter 4). Consider this example:

*Retention of qualification. Retention of qualification shall be in effect as long as three lots have not failed qualification in succession, no significant process changes in accordance with program plan and documentation requirements of Appendix A have been made, and at least one qualification by similarity has been performed in any 12-month, calendar-year, period.*

A minor point in this example is that a calendar-year period is not any 12-month period. Most of the requirement is reasonably clear except for the last part: "one qualification by similarity." A guess might be hazarded about its meaning (although guesses should never be made), but it should have been explained.

A related example is the following;

Resubmitted inspection lots. *Inspection lots which have been screened or reworked and resubmitted for quality conformance inspection shall contain only components which were in the original lot and shall be resubmitted once only for each inspection group (Group A and B). Resubmitted inspection lots shall be kept separate from new lots and shall be clearly identified as resubmitted inspection lots. Resubmitted inspection lots shall be randomly resampled and inspected for all failed groups using tightened inspection.*

Apart from the rather frequent use of the term "resubmitted inspection lot," there is a good deal that is confusing in this example. For instance, the words "for all failed groups" near the end suggests that there are a number of groups. But earlier it is made clear that there are only two groups, A and B. Then *all* inspection sampling is random; otherwise, all the statistical protection is lost. The emphasis on the separation of resubmitted and new lots does seem to be a little overdone. The probable intention can be determined with some thought, but consider the amount of time that has to be expended.

MORE LOST COSTS.

e. Accuracy and placement are important, as witnessed in this example:

Performance characteristics. *Performance characteristics shall be as specified in paragraphs 4.3.1 and 4.3.2.*

Apart from the relatively minor point that "shall be as specified" should be "are specified," the referenced paragraphs did not mention performance characteristics. They were in another specification, the component detail specification, which is where they should have been, not in a general specification, which is where the extract came from. Again I mention repetition—this time of the words "performance characteristics." This fault is not too serious except for the fact that it is time wasting and that means more LOST COSTS.

f. Concealed information is often a problem.

Change of product or process. *After qualification . . . the production and quality assurance personnel shall not implement any changes in design, materials or processes or controls without concurrent change in the process documentation and for major changes which may effect performance, quality, reliability, without concurrent notification to the program manager. The intent of this requirement is not to prevent the [supplier] from improving his product but to ensure that notification of changes which may affect performance, quality or reliability of deliverable product.*

Such notification shall include; any basic changes in design topography, passivation, metallization technique, type of bonding or package structure or materials. In addition changes which do not affect interchangeability will not require requalification.

In a later paragraph:

Changes in design materials or processing. *Records shall cover initial documentation and all changes with the date upon which each change . . . becomes effective for components intended to be submitted for quality conformance inspection under this specification, the documents authorizing and implementing the change, and the identification of the first production and/or quality conformance lot(s) within which product incorporating the change is included.*

Somewhere in the first paragraph of this example there should have been a statement making it clear that requalification was required in certain circumstances. This is necessary because the second paragraph makes it clear that for certain changes, requalification is *not* necessary. The requalification requirement is well concealed. In the case of the third paragraph it is considered that another item which the records ought to include as an essential item is the fact that requalification *was* gained after a major change. (It may be of interest that this specification was prepared by a supplier company and related to its own product.) The sentence construction is not too good either.

g. The next group of examples can fairly be said to exhibit a certain lack of clarity.

*Where this specification is referred to . . . the requirements of this specification shall be deemed mandatory, except where otherwise specified . . . or where written agreement to depart from the requirements has been obtained.*

Next paragraph:

*Notwithstanding the exception quoted [in the previous paragraph] where this specification is referred to . . . the requirements in [a later paragraph] of this specification shall apply instead of any other statement for finish.*

Not much can be said about that example.

h.   Here is another one, a definition for an inspection lot.

*Each inspection lot shall consist of a single type . . . manufactured on the same production line . . . and within the same period not exceeding six weeks.*

In a later paragraph:

Lot size. *The qualification inspection lot shall be chosen and the lot and each sublot shall contain at least twice the number of micro circuits required for qualification.*

The first paragraph suggests that the lot size could be a quite variable quantity within wide limits. The second paragraph very precisely stipulates the lot size by relating it to the actual numbers required for qualification.

i.   Another definition.

Percent defective allowable (PDA). *Percent defective allowable is the maximum percent observed defective which will permit the lot to be accepted after the specified 100% test.*

A later paragraph:

Burn in.   . . . *When a PDA is specified, if the verified failures should exceed twice the specified PDA, the entire lot shall be rejected from consideration for device classes requiring burn in and the lot may not be resubmitted for burn in.*

A later paragraph:

Lots resubmitted for burn in. *Unless otherwise specified, lots may be resubmitted for burn in one time only and may be resubmit-*

*ted only when the observed percentage of defectives does not exceed twice the specified PDA.*

There seem to be so many inconsistencies in that example that the only thing that can be said is: Confusion worse confounded.

j.  The next example is not an easy one to understand.

Screening procedure. *Screening of monolithic microcircuits shall be conducted as described . . . in the sequence shown except that, where variations in sequence are specifically allowed herein, unless otherwise specified.*

One feels that a shake of the head is the only response to that one.

k.  Here is another:

*Where end point or post test measurements are required as part of any given test method employed in the screening procedure and where such posttest methods are duplicated in the interim or final electrical tests which follow, such measurements need not be duplicated and need only be performed as part of the interim or final electrical tests. Final electrical test requirements which are duplicated in interim electrical tests shall be conducted at all interim and final measurement points as indicated.*

There has to be an easier way to say what is meant. One wonders just what it really does mean.

## 2.2   Repetition

Inspection lot identification code. [*Components*] *shall be marked by a code indicating the date on which the lot was submitted for screening and inspection.*

A later paragraph:

*. . . the date of submission for the initial inspection shall be recorded as the inspection lot identification code.*

One could say, perhaps, that two different things were being defined, although this might be too fine a point. In the first place the inspection lot identification code is the date. In the second place, the date of submission is the lot identification code. As a matter of fact, these two dates are quite likely to be different, as screening could take place some time before the following inspection.

The following example illustrates repetition in successive paragraphs:

> *... shall qualify individual [component] types or groups of related [components] ... by subjecting them to, and demonstrating that they satisfy all Group A, B and C requirements.*

Next paragraph:

> *For qualification ... shall be subjected to the inspection specified in Groups A, B and C.*

The next example must be considered to be quite exceptional, although it really did occur. It relates to a series of instructions for sampling inspection which were contained in 12 paragraphs in a complex 44-page specification. The paragraphs were not consecutive, although they are given in the order in which they appeared. One possible explanation for this unique example may lie in the fact that this specification was written by three authors. It is fairly clear that they did not write very cooperatively. In most cases the paragraphs are given in full. In one or two, only the part related to sampling instructions has been given.

1. Sampling. *Statistical sampling for qualification and quality conformance inspection shall be in accordance with the sampling procedures of Appendix B of this specification, the L.T.P.D. values and acceptance numbers specified herein, and the detail drawing or specification as applicable.*

2. Sample. *The number of microcircuits to be tested shall be chosen (independent of lot size) and shall be adequate to demonstrate conformance to the tightened inspection criteria for each of the subgroups A, B and C inspection.*

3. Selection of sample. *Initial samples (and added samples when applicable) shall be representative of and selected from the qualification inspection lot. After a test has started ... may add an additional quantity to the initial sample but this may be done once only for any subgroup, and the added samples shall be subjected to all tests within the subgroup. The total samples (initial and added) shall determine*

*the new acceptance number. The total defectiveness of the initial and second sample shall be additive and shall comply with the specified tightened inspection L.T.P.D.*

4. Group C sample selection. *Samples for subgroups in Group C shall be chosen at random from any lot which is submitted to and passes Group A tests for quality conformance inspection during the week in which the first lot is submitted in each specified Group C inspection period. Samples from the lot shall be submitted to Group C inspection whether or not the specific lot has passed the Group B quality conformance inspection. . . . when none of the lots passing Group A, during the week in which the first lot is submitted, contain the type(s) which is due to be tested, the samples for Group C inspection shall be chosen from those types in the lot being tested which have not been used for the longest time for Group C inspection.*

5. Selection of samples. *Samples shall be randomly selected from the inspection lot and inspection sublots. For continuous production . . . at this option, may select samples in a regular periodic manner during manufacture provided the lot meets the formulation of lot requirements.*

6. Single lot sampling method. *Quality conformance inspection (sample sizes and number of observed defectives) shall be accumulated from a single inspection lot to demonstrate conformance to the individual subgroup criteria.*

7. Sample size. *The sample size for each subgroup shall be determined from Table B1 and shall meet the specified L.T.P.D. or lambda. . . . may, even at this option, select a sample size greater than required. However, the number of failures permitted shall not exceed the acceptance number associated with the required sample size in Tables B1 or B2.*

8. Additional sample. *. . . may add an additional quantity to the initial sample, but this may be done once only for any subgroup and the added samples shall be subjected to all the tests within the subgroup. The total sample size (initial and added sample) shall be determined by a new acceptance number selected from Tables B1 or B2.*

9. Selection of samples. *Samples for life test shall be selected at random from the inspection lot. The sample size for a 1000-*

*hour life test shall be chosen by the supplier from Tables B1 or B2 from the column under the specified lambda. The acceptance number shall be the one associated with the particular sample size chosen.*

10. Life test time and sample size. *Whenever lambda is specified the life test time shall be 1000 hours initially. Once a lot has passed the 1000-hour life test, tests with minimums of 340 hours and maximums of 2000 hours may be initiated for new lots provided that 120 days have not elapsed since a 1000-hour life test. The sample size for a life test time other than 1000 hours shall be chosen according to the relationship of inverse proportionality between test time and sample size, such that the total unit test hours accumulated (sample size × test hours) equal the amount that would have been chosen for the 1000-hour life test had it been performed. The lot shall be accepted if the number of failures at the end of the test period does not exceed the acceptance number.*

11. Additional samples. *This option shall be used once only for each submission. When this option is chosen, a new total sample size (initial plus added) shall be chosen by the manufacturer from Tables B1 or B2 from the column under the LTPD specified. A quantity of additional units sufficient to increase the sample to the newly chosen sample size shall be selected from the lot. A new acceptance number shall be determined and shall be the one associated with the new sample size chosen. The added sample shall be submitted to the same life test conditions and time periods as the initial sample. If the total number of observed defectives (initial plus added) does not exceed the acceptance number for the total sample, the lot shall be accepted. If the observed number of defectives exceeds this new acceptance number the lot shall be rejected.*

12. Sample selection. *Samples shall be randomly accepted from the assembled inspection lot after specified screening requirements of this procedure have been satisfactorily completed. In Table C2 (Group B) for all subgroups; and in Table C4 (Group C) subgroup 3; "salt atmosphere", devices may be used that are electrical rejects when endpoint measurements are not required. In Table 1 (Group A) "electrical", a single*

*sample may be used for all subgroup testing. In Table C4
(Group C) devices used in subgroup 1, "Thermal and Mois-
ture Resistance," may be used in subgroup 2, "Mechanical."*

One wonders just what can be said about these numerous extracts
from a single specification. It is clear that they have been taken out
of context, but that does not really affect the instruction that each
provides. However, they all refer to the testing of a single class of
electronic component at different stages of a lengthy acceptance
procedure. One would have thought that, whatever the reason, if
repetition were considered necessary, the same procedural methods would be called up each time. As nearly as can be determined,
at least eight of these paragraphs deal with the selection of samples. Also, as nearly as can be determined, each describes a different method.

Four of the paragraphs either deal with, or refer to, additional
samples, and the methods described are contradictory. At the
same time, for these additional samples, there appears to be a
further contradiction. In one case, in paragraphs 3 and 8, it says
"once only for each subgroup." Then, in paragraph 11 it says,
"once only for each submission." In paragraph 8 there is another
kind of confusion, relating to acceptance number and sample size.
It says that the total sample size "shall be determined by a new
acceptance number." Ordinarily, the acceptance number is determined by a combination of lot size, sample size, and a few related
factors. In any event, the correct procedure for determinng the acceptance number is given in paragraph 9.

An analysis of these 12 paragraphs by apparent purpose is as
follows:

| | |
|---|---|
| Sample | 1 |
| Sampling | 1 |
| Sample size | 1 |
| Group C sample selection | 1 |
| Single-lot sampling method | 1 |
| Life test time and sampling method | 1 |
| Additional samples | 2 |
| Selection of samples | 4 |

If one looks at them a little more closely the following will also be seen:

1. Part of paragraph 3 is contradicted by part of paragraph 8.
2. The first part of paragraph 4 is contradicted by the first part of paragraph 12.
3. Paragraph 10 is a much expanded part of paragraph 9.
4. Paragraph 8 is repeated in bits, in a different order, and with more information in paragraph 11. The options are contradictory.

The grammatical and punctuation errors are as in the text from which they have been quoted.

## 2.3 Contradiction

a. An early sentence in a paragraph says:

Electrically and structurally similar microcircuits. *Microcircuits that are designed to perform the same type of basic circuit function using the same basic circuit element configuration and differ only in the number of identically specified circuits which they contain....*

The last sentence in the same paragraph says:

Electrically and structurally similar microcircuits shall be as specified in the detail specification.

b. With regard to a date code, a paragraph says:

Date code. *The date code shall be a four-digit number, the first two numbers shall be the last two digits of the year, and the third and fourth shall be the calendar week.*

The next sentence says:

*The date code for a given production lot shall be the calendar week in which the final seal is applied. A three-digit code may be used where required by package size.*

These two sentences seem to be a trifle contradictory as between the suggested requirements for a four- or three-digit code. The wording, generally, is rather free, as evidenced by the use of the words "digit" and "number."It should also be made clear that it is the "calendar week in the year" that is to be used.

However, a later paragraph has this to say:

Traceability. *The main purpose of traceability is to facilitate the identification of potential defects wherever a failure mode is suspected. All devices are marked with a four-digit date code that identifies the week and year of manufacture and provides traceability to the seal date code.*

Here there is no question of a three-digit date code, only a four-digit code. There is also a slight vagueness about the purpose of traceability.

c.    Another slightly confusing example of contradictions:

Process conditioning. *100% process conditioning shall . . . be in accordance with paragraphs 4.5.1.1 through 4.5.1.8 as required by Table 1.*

It was noted that there was no Table 1 in the specification. However, a later paragraph said:

Processing options. *When specified in the Purchase Order, the tests referred to in paragraphs 4.7 through 4.11 shall be performed on each component. These are processing options which will be identified on the purchase order by "slash" numbers.*

A following paragraph said:

Process option 1. *Shock series for package integrity, including thermal shock, paragraph 4.5.1.1; mechanical shock, paragraph 4.5.1.6; fine leak, paragraph 4.5.1.8; gross leak, paragraph 4.5.1.9.*

It is a little difficult to determine whether this example is one of repetition or contradiction. In any event, according to the second paragraph, the processing options are described (in 4.7 through 4.11). But according to the third paragraph, the options are described in 4.5. . . .

d.    In this example the first statement is nearer the truth than was probably intended.

Conflicting requirements. *In the event of a conflict between the requirements of this specification and other specifications referenced herein, the precedence in which requirements shall govern, in descending order, is as follows:*
*(1) Applicable device specification.*
*(2) This specification.*

In a later paragraph the following statement appears:

*The detail specification . . . shall be prepared in such a format and content, however, as to clearly identify which parameters, limits and conditions of test shall be exercised in response to any given requirement of this specification.*

e. This example is a rather subtle demonstration of the art of contradiction without the appearance:

*. . . shall establish and maintain inspection at appropriately located points in the manufacturing process in accordance with the procedure described . . . to assure continuous control of quality of materials, subunits and parts during fabrication and testing. This inspection shall be adequate to ensure conformance with the applicable detail specification.*

One would want to know whether the inspection is to be in accordance with the "procedures described" or the "applicable detail specification.

## 2.4 Nonsense Statements

a. The first example in this section seems to be one in which the writer was carried away by his or her own eloquence:

*In order to obviate the possibility of subsequent modification and the high expense involved, the vendor is requested to make every endeavor to interpret the requirements of the specification in such a manner that the primary concept is correct.*

It was noted that nowhere in the specification was there any reference to the "primary concept"—or to any other concept, for that matter. It should also be noted that the reader of the specification is "requested . . . to interpret" the requirement. This task—*interpretation*—should never have to be undertaken with regard to a specification. However, the next sentence in that specification went on to say:

*If a particular detail of this specification is not clear or is capable of more than one interpretation, then the vendor must request clarification from [the customer].*

This is also an excellent example of contradiction, as the second sentence specifically prohibits that which the first sentence urges the reader to do: that is, "to interpret."

b. The next example purports to be a definition of rework. Most people who work in industry will have a pretty clear idea of the meaning of the term. This definition is likely to upset those ideas. Attempts have been made to rewrite the example into "normal" English, but without success.

Rework. *Any processing or reprocessing operation, other than testing, applied to an individual device, or part thereof, and performed subsequently to the prescribed nonrepairing operations in the manufacturing process which are applicable to all devices of that type at that stage.*

It might be said that nowhere in the specification could any reference be found to "nonrepairing" manufacturing operations, prescribed or not. In any event it was felt that if the definition meant what it seemed to mean then, rework could not be carried out.

c. This little example might be thought to be "discreet":

*The detail specification shall include:*
*a) part number and number of the applicable detail specification.*

d. In this example, in which the product is required to conform to a *very* unusual set of requirements, a real gem has been unearthed:

Qualification. *All devices furnished under this specification shall be products which have been tested and have passed the following qualification requirements:*
*1. Manufacturer's understanding and knowledge of the product and awareness of expected manufacturing reliability problems.*

2. *Availability at the manufacturer's plant of the specialist equipment necessary for the manufacture and control of the product.*
3. *Manufacturer's general understanding of reliability as demonstrated by an in-house program.*
4. *Data demonstrating ability to meet the requirements of this specification.*

It has to be said that it would be a very strange product indeed which could pass these peculiar qualification requirements. I have never heard of anyone, anywhere, requiring a product to "pass" such peculiar (for this purpose) parameters. As a matter of fact, they are very much the kind that one would expect to be applied to a manufacturer, one being assessed regarding the suitability of his or her quality control system, perhaps in accordance with the requirements of the specifications discussed in section 3.5 in Chapter 6.

e. This statement was printed on an errata slip in a government publication:

*Any reference in this document to a week should be taken to mean a fortnight.*

Any comment on that one is surely unnecessary.

f. Another unclear definition is the following:

Index point. *The index point indicating the starting point for numbering the leads shall be as specified. The indexing point may be a tab, color dot, or other suitable indicator.*

The index point is to be "as specified." But it may be this, it may be that, or it may be anything suitable. It might have been simpler to have said something like the following:

Index point. *Any suitable indicator may be used to identify the starting point for numbering the leads.*

One might also suggest that there could be a subtle difference between an "index point" and an "indexing point."

g. Try this one for size:

*However, all tests, preconditions, and screening operations which were performed on all components submitted for qualification inspection shall be performed on all components subsequently sub-*

*sequently submitted for quality conformance inspection until the next qualification inspection is completed and passed to assure that qualification inspection samples are representative of components being offered for acceptance.*

There are two levels of inspection in the system being described here. The first, which may be called a qualification approval/ inspection, is an initial inspection prior to formal approval for the commencement of production deliveries. It is usually much more comprehensive and severe than the second, routine, lot-by-lot, "quality conformance" inspection, which is carried out for normal production acceptance and delivery purposes. It would be quite unusual to insist that "qualification inspection" should be carried out on every lot from the time of initial qualification until the next qualification inspection was due. I have never come across a case in which this was even suggested, let alone made a requirement.

The real meaning that the requirement is intended to convey is, to say the least, rather hard to determine. It is quite likely that the writer did not mean what he or she actually said. It is possible that what the writer did mean was that the quality conformance inspection is also carried out on the qualification approval lot. That would not be out of the ordinary. But again, trying to understand the meaning of this requirement would incur more LOST COSTS.

## 2.5 "Dangerous" Extracts from Other Documents

Making use of extracts from another specification, a national standard, or any other kind of document is quite common and there are many reasons why it is done. Basically, there are two ways in which this process of extracting information can be carried out. In the first place, with or without acknowledgment of the source, the required extract is given verbatim. In the second place, the source is clearly identified and the appropriate references made to a particular page, section, paragraph, and so on. Which of these two methods is used is likely to be a matter of personal preference. There are pros and cons for each.

It should, however, be borne in mind that if an extract is made from any other source—for example, national standard—without identifying the source, problems could arise in the future: one immediate and the other long term. The immediate problem can arise if an error is made when transcribing the extract and the source is not given. If they suspect an error, readers can only go back to the originator for clarification—that will not always be done. Thus the error will remain.

The second, longer-term problem will arise if subsequent changes are made in the source document. Changes that might affect the extract. Experience has shown that such changes often go unnoticed, except in very well organized operations, so that both the writer and the users are left in ignorance of the fact that a change has taken place in the source document. Of course, the change may not matter. But who is to know?

Such problems can largely be avoided if the second method of making extracts is used. The writer does not make verbatim quotes, but gives the precise text references in the document concerned. It is up to the reader to obtain the source document. In this way it is much less likely that errors will arise. The problem may not be completely eliminated, but it should be greatly reduced.

As, inevitably, specification writers will quote extracts from source documents without identifying them or even giving any indication that they are extracts, attention should be drawn to another problem that can arise. This happens when an extract is taken from a source document that the writer does not fully understand but, for some reason, feels must be included. It may also happen that the writer's source took it from a tertiary document, also without understanding it. Readers may be assured that this does happen, and an example will be given shortly. There is only one sure line of action in such a case: *Never make use of an extract from another source unless you fully understand it.* You can be pretty certain that if you do not understand it, your readers will not either.

a. An excellent example of this kind was found in a footnote to a table of hypergeometric sampling plans for small batches. The table was given in full and there were four footnotes. The last of these read as follows:

*The sequence of sample sizes and lots are generated by taking the products of preceding numbers in each of the respective sequences and the numbers 2 and 5.*

One can discern a few puzzles in this note. First, what does one do with the *first* number in each of the respective sequences? These numbers do not have a "preceding number." Second, surely the product of a number and 2 and 5 is the same as the product of that number and 10. There is also the minor point that the secret *is* generated (not *are*).

I have not been able to find any explanation for this footnote, not even from those who are knowledgeable on the subject of hypergeometric sampling plans. It seems clear that the writer of the specification must have taken the table, complete with footnotes, from another source, although he or she could not have understood it. A year or so ago I found the missing source—an American Military Specification. As the table complete with its footnotes is in the Mil Spec, it also seems to be a reasonable assumption that the writer of the Mil Spec had done the same thing—extracted information from another source without understanding it. I am awaiting the day when someone, perhaps a reader of this book, will draw my attention to the tertiary source. Perhaps it will be a statistical textbook in which will be found the table and the correct footnote.

It would be interesting to know why it was thought necessary, in any case, to include the table and its footnotes. Apart from this reference, I had never in 40 years of work in industrial quality control heard of these sampling plans or of anyone using them. They must have a purpose—or do they?

b. Another example concerns an engineer who was preparing an in-house specification for sheet steel for deep-drawing purposes. In the specification he called up a British Standard for sheet steel which could be used for deep drawing. In due course sheet steel was ordered in accordance with the in-house specification for deep-drawing steel. Later, after receipt, when the press shop attempted to use the steel, the sheet was found to be quite unsuitable for the purpose. When an investigation was made, it was found that a note in the British Standard said:

*When steel to this standard is ordered for deep-drawing purposes, this fact must be stated on the order.*

This special note had been overlooked by the engineer. Or perhaps he had not even noticed it. Which again raises the point that it is undesirable to qualify statements that have already been made. Remember the "qualification" from Leviticus 11:3 that appeared in Chapter 1?

Specifications, standards, and procedures should always be

read with care. It is unfortunate that this is not always done, so errors arise.

## 2.6 Excessive Requirements

There are two variations of this fault. The first relates to cases in which later sections of a specification call up more stringent requirements than have already been called up. The second relates to the case in which the writer of a specification has not taken into account the consequences of a specific requirement that he or she has written into the specification.

Although it may not always be easy to see that it is so, some of the examples given earlier in this chapter illustrate the first kind. For instance, examples 2.1f, 2.1j, 2.2c, and 2.4a all show that a tightening up of requirements has taken place. In example 2.2c, additionally, it may seem as though one was going first one way and then the other.

However, by far the best example of the second kind that I have ever seen is that of a reliability requirement that appeared in a detail specification for transistors. It concerned the long-term life requirements and was expressed more or less in the following terms:

*The manufacturer shall demonstrate a failure rate which shall not be worse than 0.0001% per thousand hours.*

However to "demonstrate" a failure rate of less than 0.0001% per 1000 hours, the manufacturer would have been required to carry out a quite extraordinary exercise. He would have been required to place on test no fewer than 2000 transistors for a period considerably in excess of 100 years without experiencing a single failure. Quite apart from the general impossibility of such a requirement, and the astronomical costs (more LOST COSTS), one wonders who, aside from the customer, could possibly be interested in the final outcome of such an exercise.

## 2.7 Multiple Mixed Specification Systems

Fortunately, this is not a very common problem area and is usually found only in specification systems developed by the larger organizations. However, when it does occur, it can present many headaches for those who have to deal with it—usually people in smaller companies. In the example described below, a series of four American specifications was involved. They were the basis of an order from a customer in Paris to a supplier in West Germany. (That alone begins to make for confusion.) The original specifications were written in English and translated into German.

The four specifications consisted of one detail specification for the part itself and three general, and quite complex, environmental and reliability specifications. The three dealt with a wide variety of requirements and averaged 40 pages each. The four specifications totaled about 130 pages. Quite a lot to have to consider for what was actually a very small component, a transistor.

The detail specification contained a complex series of acceptance tests to be carried out. Each test was linked to one or more of the variety of environmental and reliability requirements: for example, electrical testing at high and low temperatures and with vibration applied. Because of the sequencing of the electrical tests and the ordering of the other requirements the latter were called up in an apparently quite random manner. It was necessary to skip from one place to another through all three of the general specifications.

A further complication was the fact that the group of inspectors who would be responsible for the acceptance testing consisted of eight young women of six different nationalities from Turkish to American! They did not have a fully common language, although all had at least some understanding of either German or English. It must be admitted that this was not exactly a typical situation, but it did happen and these specifications did not make matters any easier.

The relevant parts of the three general specifications were extracted in the order in which they were required by the sequencing of the electrical tests. This was by no means in any consecutive order. The electrical tests, together with their linked environmental and reliability requirements, were then collated into a new document. In this, the tests were all in the correct sequence and relationship. This new document amounted to about 40 pages instead of the 130 of the originals.

The new document was then translated into German and the two versions, English and German, were issued to the inspectors. They then had the assurance that, voluminous though it might seem, they now had only the essential information *in the order in which it was required*, and without some 90 pages of useless material.

The entire exercise took about six weeks, as many problems arose which were caused by the usage of words in the original specifications. In many instances clarification had to be sought from the United States because the usage of words did not always conform to normal English usage—or to German usage either. In at least one instance the "clarified" meaning was in complete conflict with the normal meaning of the words that had been used. It was not possible to get a reason for this apparent contradiction.

When a number of more or less complex specifications have to be used in parallel in cases such as this, the originating organization has a problem. It is very tempting to consider that the compilation and presentation should be done in a manner which best suits their own convenience. But it is not always appreciated that this may create considerable problems for smaller companies, as in this case. Not only may they be smaller, but the costs incurred in overcoming the problems are, in the aggregate, likely to be much higher than the original cost savings—more LOST COSTS.

## 2.8 Chain Specification Systems

This represents another very frustrating side of the use of specifications to those users who have to examine them in detail— again, especially in the smaller company. This is the situation in which the first specification calls up a second, the second a third, the third calls up a fourth, and so on. The frustration of readers increases, as does their disgust, especially when, as is sometimes the case, the dependent specifications are not supplied in the first place. Users find that they have to ask for them as they realize that they need them, although quite often the need is more apparent than real. The example described here not only highlights the problem, but must be considered a classic of its type.

A supplier company was invited to tender for the supply of a group of electronic components. Together with the invitation to tender, the supplier was sent a very complex specification 30 pages long (It is identified here as specification B.) It was carefully read by the supplier, who concluded from his reading that he also required specifications A and C which had not been sent to him. It happened that he had a copy of A but had to ask the customer for C.

In due course C arrived and was carefully read in conjunction with A. The conclusion was that, while some of the references to A were relevant C was not relevant in itself. However, in both A and C there were references to D which suggested that it, too was relevant. Of course, it had not been supplied, so it was requested. In due course D arrived and was read as carefully as had been its predecessors. It was not relevant but—surprise, surprise—there were references to two further specifications, E and F, which appeared to be relevant. It might also be mentioned that D referred to a total of 50 other specifications, drawings, and other documents.

E and F were requested and arrived in due course, and also carefully read. By this time it was no surprise to realize that neither of them was relevant, but there were references to yet another specification, G, which suggested that it, too, was relevant, and also to A.

Although G was requested, it never arrived, and the supplier gave up. In a way it was rather a pity that he did so, as one will now never know if G might not have called up H, I, J, and so on ad infinitum. The time scale from receipt of the orginal invitation to tender, to the request for G was about six weeks. In this period much valuable time was wasted.

More LOST COSTS.

There are at least two main lessons to be drawn from this example, primarily for specification writers with an involvement in specification systems. The first one should, in fact, apply to all specification writers: *Never make reference to another specification unless it is absolutely necessary and truly relevant; unless it is necessary for the reader to refer to it; unless the precise, relevant part of the referenced specification is fully identified.* The second lesson should be: *Check all references and cross-references to make sure that the reader will not be involved in superconducting circles of the type described above.* This type of quality circle does not serve any useful purpose—quite the reverse. Perhaps there is a third lesson to be drawn from that example, although one is not quite sure for whom it is intended. *When there are consecutive references from specification to specification, care should be taken to ensure that all the relevant specifications are sent to the intended reader at the same time.*

## 2.9 Multiple Authors

There is an old saying that a camel is a horse designed by a committee. This is not to say that responsibility for the preparation of a specification should not be given to a committee. However, the committee should ensure by all possible means that the writing of the actual specification, standard or procedure is entrusted to a single writer. If, for any reason it is considered necessary to use more than one writer, the committee must also ensure that they create a coordinating function. A principal task of this function must be to ensure that there is no contradiction, no repetition, no confusion, and so on. Experience has shown that in the circumstances in which there is more than one writer, there is often a lack of consultation between them. Inadequate briefing can also result in overlapping content. Example 2.2c is a remarkable case of what happens when multiple authors do not fully collaborate.

## 2.10 Omissions

In this group would be placed examples in which, say, referenced tables are not to be found. It is also intended to cover other, less obvious omissions, such as cases in which a specification writer should have included an item of information but has not done so. The missing item is often an adjective or description of something that has been said, such as those given in example 1.4k in Chapter 4, and in section 1.4a. One other good example of "missing" information is that of example 2.1f in Chapter 5. Others will be found scattered within those chapters.

# 6

# Specification Faults:
# Group 3—Presentation

No two persons are likely to form the same opinion about the relative importance of the various classes of faults. I am inclined to believe that presentation, the class under discussion here, is perhaps the most important because feelings about a specification are likely to be aroused by the way it looks at first sight. There may be no other faults present, or there may be many, but if the presentation is poor, a wrong impression may be created, an impression that other faults must be present. Readers may well feel that if the writer is appparently not bothered about how they will receive the material, why should they bother either? Or readers may feel that the writer was merely careless—with the same result. Such a specification will be less well accepted and, more important, more likely to be disregarded. Readers may draw conclusions without sufficient examination of the detail because they have been put off by the poor presentation.

## 3.1  Lack of Consideration for the User

One type of lack of consideration is to assume that the user has a higher level of knowledge than is the case, or perhaps, even likely. Here is one such example:

a. Before dice are approved for assembly, they are to be checked

under high power microscope for cracks, chips, oxide deficien-
cies and improper aluminum interconnects. . . .

It is unlikely that the users will know the order of magnification
that is required. Nor could they find the information anywhere else
in the specification. It is information which cannot, must not, be
guessed at, and it is important.

   b.   In an example I dealt with some years ago, one requirement
in a processing specification was given in one 10-line paragraph.
As written, this requirement was so complex that the production
personnel could not understand it. It was necessary to rewrite it in
a greatly expanded form to put it at a language level understandable by the production personnel. The revised version was 1½
pages long. The original requirement had been written by two engineers who were very knowledgeable about the subject. They did
not stop to consider that the production personnel were not as
knowledgeable and would not be able to cope with their mental
and verbal shorthand.

   c.   Another example concerns a document that was intended to
be a guide to certain statistical techniques for those with little real
knowledge of the subject. It was about 100 pages long, but by the
time readers got about as far as page 10, they found it heavy going.
And the further they got, the more complex the material became—
which, of course, defeated the object.

   d.   A final example in this section is taken from the 1978 British
Consumer Safety Act. It is a definition for goods:

Goods includes . . . :
a)   *In relation to a Notice to Warn, includes things comprised in
     land which by operation of the law become land on becom-
     ing so comprised.*

When I first came across this definition, I asked the company
lawyer for an explanation. He said that it was a question for which
he "required notice." He produced an answer three weeks later. It
concerns the difficult question of who owns what when the owner
sells a house on which many improvements have been carried out,
some of which are more permanent than others. It is doubtful that
anyone other than a lawyer could understand its meaning, although it is said that ignorance of the law is no excuse.

## 3.2   Unbelievable Examples

One of the best ways of conveying information in specifications, standards, procedures, instruction manuals, and books like this is to provide practical examples. Whenever I have had a need for practical examples in training courses, I have always done my best to ensure that they are as realistic as possible and, most important, that they are believable. It is pleasing when, during a course, participants say: "Why, this is just like my company! Who told you?" Unfortunately, not all examples given in specifications are believable, and when they are not, readers' opinions of the credibility of the specification drop. Although perhaps not entirely justifiably, I feel that way about examples involving that mythical object, the widget.

a.   Following is an example related to the production of small, core laminations for transformers.

*The production rate is 1200 per hour.*
*The defect rate is 6%.*

Anyone with any knowledge of press operations on small parts like this would expect a production rate on the order of tens of thousands per hour. And, any self-respecting press shop foreman would be tearing his hair out if the reject rate was as high as 0.1%, let alone 6%.

b.   A similar example concerned the production of a small brass pressing produced at the rate of 4 to 4½ thousand per hour:

*There is a depth tolerance of 0.1 mm gross and control limits were determined at 7.7 for high and low limits and 12.7 for total number of defectives.*

The first comment is much the same as for the preceding example. For a small component of the type described, with a fairly generous tolerance of 0.1 mm, one would hardly expect a defect rate of the order suggested by the control llimits—this would appear to be in excess of 5%. In any event, the actual numbers quoted do not seem to make sense.

c.   Next is an example related to an automatic process for the manufacture and weighing of small pellets.

*The production rate is about 750 per hour. The weight is 155 gr. and the weight tolerance is plus or minus 1.5 gr. Two small indicating scales are used for check weighing. One for the high tolerance and another for the low tolerance limit. . . .*

Two scales were required for the following reason:

*. . . since it is only necessary to see if a pellet falls outside the weight tolerance (with two scales) checking could be done more rapidly.*

A singular need is that requirement for two indicating scales. One ought to be sufficient to indicate if a pellet was outside either limit, and quicker, too.

More LOST COSTS.

### 3.3   Untrue Statements

Statements that are untrue, whether obviously or not, further help to reduce the degree of credibility that readers will have for a specification. If the statement is complex, that will make matters worse, as readers will be even more confused.

    a.   Consider this example:

*Any measuring instrument (and this term includes such things like standard resistors and volumetric glassware) must at some time in its career have been adjusted or graduated so that, within appropriate limits, it tells the truth.*

The wording in this example is rather unusual, to say the least. However, if it is considered carefully, the phrase "must at some time in its career" will stand out. It is not exactly a true statement, apart from the rather careless choice of words. "Career" is a very odd word to use when discussing the life of inanimate objects, or even animals. However, except for items such as volumetric glassware which once calibrated are unlikely ever to be recalibrated, most measuring instruments are adjusted many times during their working life. The words "must at some time" strongly suggest a "one time only situation, and that is not so.

    Consider the next sentence in this example:

*For all measuring instruments this will have occurred as one of the final steps of the manufacturing process. . . .*

There cannot be many measuring instruments that are not adjusted or graduated as the final step of the manufacturing process.

b.   In a guide there was a reference which said:

*. . . a level of 0.33 defectives [was set] on the upper control chart with a level of 0.67 defectives on the lower chart.*

The chart referred to did not have levels of 0.33 or 0.67 defectives indicated.

## 3.4   Philosophizing

The following was a proposed definition:

*Risk quantification cannot measure risk acceptability, which is the concern of the relative decision makers who must judge the benefits, alternate use of resources, and other factors unrelated to the process of risk quantification. Moreover, the uncertainties in the quantification do not become entangled quantitively in the process of judgment. They are simply determinants of the potential range (and perhaps distribution) or variation of the values of the quantified risk. This range can be assessed for acceptability.*

This statement appears to be a case of philosophizing at a fairly high level and should not be a part of any specification—apart from the fact that this definition is difficult for most people to understand. Incidentally, it was changed to the following:

*The estimation of a given risk by a statistical and/or analytical modeling process.*

Example 2.4a in Chapter 5 is also a suitable candidate for inclusion in this category.

## 3.5   Reading Across and Illogical Ordering of Contents in a Specification System

This is a problem that is seldom found, but when it does arise, is rather irritating. It can be found when there is a series of specifications dealing with an integrated and progressive system of requirements. One example is the series of British Defence Standards 05-21 to 05-29. Collectively they describe a progressive system of quality control from simple inspection to a full assurance system, including design control. As might be expected in such a progression, the requirements at the lowest level reappear, with additions, at the middle level. Then those in the middle level, with further additions, are found again at the highest level. In the case of many companies wishing to gain access to such a system, it would be reasonable to expect them to seek entry to the progression at a lower level and, with time, move up to higher levels.

To ease such progress, and in any case to make for easier understanding of the complete system, it is of considerable value to be able to read across from the lowest-level standard to the highest. Unfortunately, because of the ordering of the contents in each standard, this was not possible. The task of seeking a general understanding was made more difficult. About one year after the issue of these Defence Standards, the British Standards Institution (BSI) issued what might have been called the general industrial equivalent of the Defence Standards. This was BS 5179 and followed the same pattern of ordering of the contents.

In addition, in some cases, the titles of similar sections in the different levels were changed, although there did not seem to be any special reason for this. BS5179 was proposed for revision at least twice. For the second revision the requirement titles and ordering for Parts 3 and 2 were as given in Table 1. Where a title in Part 2 differs from its equivalent in Part 3, it is given in full. It will also be noted that there are two requirements which, although in Part 2, do not appear in Part 3. This seems to be an odd omission. (Part 3 is the highest level.)

Of the Part 3 items that are not in Part 2, the only omission that seems fully justified is design control. That is not a requirement for a Part 2 system level. Also, the Part 3 requirements of sections 3, 4, 5, 12 and 16 ought to have been included in the Part 2 requirements. In neither case was the ordering very logical. A more logical ordering for both parts might be as follows. Of course, item 4, design control, would appear in Part 3 only.

1. Organization
2. Periodic review of the system
3. Planning
4. Design control
5. Documentation and change control
6. Work instructions

**Table 1**

| Part 3 | Section | | Part 2 |
|---|---|---|---|
| Organization | 1 | 1 | (same title) |
| Periodic review of Q.A. system | 2 | 2 | Periodic review of inspection system |
| Planning | 3 | | |
| Work instructions | 4 | | |
| Records | 5 | | |
| Corrective action | 6 | 13 | (same title) |
| Design control | 7 | | |
| Document and change control | 8 | 3 | Documentation |
| Control of inspection and measuring and test equipment | 9 | 4 | (same title) |
| Control of supplier procured supplies and services | 10 | 5 | Control of purchased supplies and services |
| Manufacturing control | 11 | 6 | In process control |
| Purchaser supplied items | 12 | | |
| Completed item and test | 13 | 11 | Completed item inspection |
| Sampling procedure | 14 | 9 | (same title) |
| Control of non-conforming material | 15 | 10 | Non-conforming items |
| Indication of inspection status | 16 | | |
| Handling storage and delivery | 17 | 8 | Handling and storage |
| | | 12 | Packaging, marking and delivery |
| Training | 18 | 14 | (same title) |
| | | 7 | Workmanship |

7.   Control of inspection and measuring and test equipment
8.   Control of supplier procured supplies and services
9.   Purchaser supplied items
10.  Manufacturing control
11.  Sampling procedures
12.  Completed item inspection and test
13.  Control of non-conforming items
14.  Indication of inspection status
15.  Corrective action
16.  Handling, storage and delivery
17.  Records
18.  Training

With the replacement of BS 5179 by BS 5750, there has been some improvement in the "reading across" aspect, although it is still not as good as it might be. However, the requirement titles have been unified for all three parts, and most of those omitted from Part 2 of BS 5179 are now in Part 2 of BS 5750.

Of course, a principal reason for requiring logical ordering of the contents of a specification is so that the writer can take readers through the subject literally from beginning to end. This makes it easier for readers and greatly helps their understanding. This is a problem primarily where there is a progressive series of specifications for generally similar purposes, although the principle ought to be applied in all cases and for the same reason. The "reading across" example of BS 5179 discussed above is also a good example of illogical ordering. But it does happen in single specifications, as the following example shows:

> *Quality conformance inspection.* Quality conformance inspection shall consist of the examinations and tests specified in paragraphs 4.3.1 and 4.3.2.

Later in the same specification is the following additional statement:

> *General inspections and tests.* General inspections and tests shall be applied to all components intended to be submitted to quality inspection or quality conformance in accordance with this specification....

As the second referenced group of inspections is clearly intended to be carried out before those first mentioned, they should have been dealt with first in the specification.

### 3.6 Poor Visual Material

Fortunately, not many examples of this type of fault have come to light, although they do occur. When they do, they seem mostly to have been caused by careless drafting during the preparation of the diagrams. At the same time, in other examples the drafting is excellent but the associated text references are incorrect, so that the diagram may not be understood. Also, of course, sometimes the diagram is simply wrong. One or two examples of this type have been given in section 1.5 in Chapter 4.

One example was found in a discussion about the differences between *low relative precision, medium relative precision, and high relative precision.* Case b is "medium"; the accompanying text said:

*It can be seen that in case b), shown in Figure 12, the defectives to*

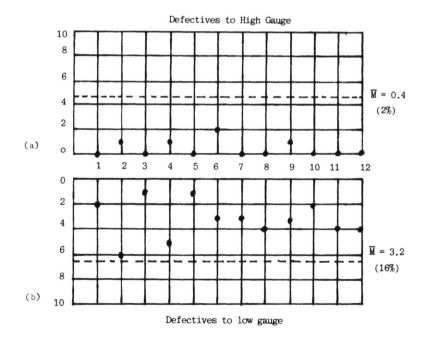

Figure 1 Example Figure 12.

*the inspection limit exceed 16% even although no rejects are being produced, the control chart indicates the drift in the average quite clearly. This enables action to be taken to control the drift before rejects are produced.*

Apart from the problem of how defectives can be found even although no rejects are being produced, an examination of the control chart does not show clearly that "the drift in the average [is shown] quite clearly." Indeed, one would be hard put to perceive any drift at all. There is much in this example that is not understood, and this adds to the confusion already caused by this poor-quality visual material.

## 3.7 Unsatisfactory Proofreading

We all know how difficult it is to eliminate errors completely. A good way to try to ensure the absence of proofreading errors is to ensure that someone other than the writer reads the final draft— preferably someone who is not familiar with the subject. When one is familiar with a subject, one tends to anticipate the text. The eye correctly passes the error signal to the brain, but the signal is ignored because of the anticipation, which is overriding.

Perhaps the greatest danger arises during the preparation of a draft document when one important item of information is not yet known. In that case, to make sure that the draft is otherwise complete, a dummy item of information is inserted in place of the as-yet-unknown item. The basis for this action is that the dummy item is so obviously incorrect that, later, when the real item becomes available, an exchange can be effected. That is the theory.

Unfortunately, this does not always happen, and the document goes for printing and distribution with the dummy item of information there for all to see—except by the final proofreader! The following example must surely rank as the most embarrassing of its kind ever to have happened. For obvious reasons, the name of the company concerned is not given.

Preparations were being made for a company-only exhibition to be held in central London. Thousands of brochures were to be sent out to customers with all the information that they could want or need to know about the company and the exhibition. One important item was that special telephone facilities were to be installed for the convenience of customers. However, when the draft of the brochure was being prepared, the special telephone number was not known. The publicity manager decided that rather than leave a blank space for the time being, and to make the text look real, he would insert a number so obviously wrong that it could not possibly be overlooked later at the final proofreading stage.

You've guessed it: That incorrect number *was* overlooked and the brochure was printed. Copies were distributed to customers all over Europe. Then, just a few days before the exhibition was due to open, I happened to see a copy of the brochure. There before my horrified eyes was that *special* telephone number, the number that was supposed to have been provided for the convenience of customers for the duration of the exhibition. The "obviously incorrect" number that the unfortunate publicity manager had used was the number which, arguably, was the best known telephone number in the country—Whitehall 1212—then the telephone number of New Scotland Yard, London's police headquarters.

It was much too late for anything practical to be done about this gross error. For a week Scotland Yard had many more wrong numbers than they had ever imagined possible. And a most abject apology had to be presented to the *police* on behalf of the company by the publicity manager. The moral of this sad, but true, tale is: *Never deliberately put incorrect information into a draft document, whatever the reason may be.* Instead, leave a blank space. The consequences will be much less serious if the blank is missed at the proofreading stage than if false information is overlooked.

## 3.8   Lack of an Index

Lack of an index is a fairly common fault and can be very irritating, especially in the larger specifications. The Defence Standards mentioned in section 3.5 of this chapter did not have an index, a fact that hindered understanding of the complete system. When reading a long and complex specification, eventually one will come across a point that seems to have relevance to earlier text. But one cannot remember where that earlier text is, as there is no index. So, much time is wasted in searching back through earlier pages. When, as is sometimes the case, this happens a number of times, it becomes a considerable nuisance. It becomes another LOST COST.

The specification that is dealt with at length as an exercise in Chapter 18 is a case in point. It did not have an index. However, there were so many problems in reading it that an index was compiled as a part of the exercise. The index is included in Chapter 18.

# 7

# Case Study 1:
# Customer Specification Problem

In this chapter we first examine the circumstances of Case Study 1 in some detail. Then we look briefly at the types of problem that supplier companies can face when customers introduce complicated specification systems

Over a period of many years a large supplier company received an average of 25 customer specifications each week. They ranged in complexity from single-sheet copies of the supplier's own data sheets to documents of 50 pages or more. The latter covered entire classes of components and were accompanied by individual detail specifications for each component. Originally, these specifications had been dealt with by the engineers in the product manufacturing departments, according to component type. As many engineers were involved, there were many degrees of care in the scrutiny to which these specifications were subjected. Perhaps not surprisingly, it was common for important points to be missed.

Most of the missed points were usually picked up at the final acceptance stage, which was somewhat inconvenient. Some were not picked up at all, which was exceedingly inconvenient. In the latter case the first indication that there something was wrong was a rejection from a customer. As a result, a good deal of time had to be spent in investigating the circumstances before the actual fault was revealed and the situation corrected—more LOST COSTS.

After a long period of such problems the quality control department decided that something had to be done. The product departments and their engineers were showing little interest in the blurring of the company's image that was taking place. Nor did they seem very concerned about taking corrective action. The proposal put to the product departments on the initiative of a senior quality engineer was to the following effect.

When the product engineers had completed their normal initial examination of the incoming customer specifications, the quality department would be given an opportunity to examine them. If any points were found that were thought to be of significance, they would be brought to the attention of the relevant product engineer. It was the intention that this information would be used by the product departments to decide three things;

1. Was the company prepared to accept the specification?
2. If the acceptance was conditional, what points would have to be taken up with the customer?
3. How best can the internal specification, prepared in the company's "language," accurately represent the requirements of the customer?

After much discussion the product departments agreed to give the proposal a try. It should be said that much of the objection to the proposal came from the individual product engineers, who were afraid that the quality department was trying to usurp their authority! It took a great deal of persuasion before they would accept that, on the contrary, the quality department was trying to do something different—to provide them with a backup service, which they had never had before, which would strengthen their hand in negotiations with customers. In addition, of course, it would improve the image of the company.

Unfortunately, the end result was not quite what the quality department had hoped for. The situation did not improve at all, not because any of the specifications were missing the quality scrutiny, but sadly, because in most instances, the product departments took no action on referrals to them. The reason given was lack of time due to the pressure of work.

After a further lengthy period of this unsatisfactory situation, the quality department decided that further action should be taken. This time a quite different proposal was put forward—one that would give the quality department principal responsibility for internal scrutiny of incoming customer specifications. The quality department would also be responsible for ensuring that any queries that arose were resolved. Based on a reasonable time scale, it was proposed that lack of, or incomplete, action on the part of product engineers would have to be justified by them. Not surprisingly, there was considerable opposition to this proposal from the product departments. But the quality department had done its homework. The strength of the evidence about customer specification problems, given to the product departments, was so strong that the product engineers could not reasonably object to the proposal. It was agreed to with reluctance.

At this point it is desirable to go back in time some years to highlight another customer specification problem for which, up to that time, no satisfactory solution had been found. The problem arose from the fact that there were so many customer specifications on file that were active. Because of this, and for other reasons, it was not possible to check every current specification for every proposed process or product change. Actually, it was possible to check only a few.

The case that highlighted the problem was the following. A customer some time earlier had sent in a specification which was actually a copy of one of the company's standard data sheets: a not uncommon occurrence. In the scanty customer records that were maintained at the time, this customer requirement was recorded as "standard." As the customer placed orders for his "special" specification, the marketing department automatically translated it as "company standard." As such, it was eventually shipped. The customer ordered this item about every six to nine months.

Over time, the product department began to consider a product cost reduction program for this component. It so happened that it had four leads, one of which was connected internally to the case. The remaining leads were connected to the three individual elements of the component, also internally. The cost reduction proposal was that one of the internal elements would be connected to the case. This would save the cost of one lead, which, as they were gold plated, was appreciable. It was thought that this change would not cause any functioning problems in circuits.

The product applications department was consulted and confirmed that, in their opinion, there should be no inconvenience to any user of the component. The change proposal was agreed to and put into effect. Instead of being a four-lead component, it was now one with three leads. No one realized, or remembered for that matter, that there was one customer that actually wanted and specified a four-lead component.

Some time after the change had been made, the customer sent in one of his periodic orders. As usual, the marketing department "translated" it into "company standard." The standard product, now with three leads, was shipped to the customer. In due course it was rejected by him as being "not to specification."

The quality department then checked the "records" for the customer specification and, of course, they said that the customer requirement was "company standard." Thus the belief that the components must be correct according to the customer's specification and had been rejected by mistake. They were therefore returned to the customer with a note to that effect. End of story (so it was thought).

Very soon the shipment was rejected by the customer for the second time. On this occasion, there was a note amplifying the rejection. It said:

*Our specification calls for a component with four leads and those which you have sent to us have only three leads! Therefore, they could not possibly be to our specification.*

Of course with that explanation, it was quite obvious what had happened. But it was not quite so obvious what, if any, corrective action could be applied in this case, or for that matter, to prevent similar cases from happening in the future.

The product department could not go back to a four-lead component, not even as a "special" for that customer. There was only one thing to do: tell the customer what had happened and offer the services of the applications department in modifying the circuit requirements so that the three-lead component could be used.

Although other situations of this type had arisen in later years, none had been quite as traumatic. But as stated earlier, no solution to this problem had been devised. Any solution had to wait until computer facilities were available.

Now let's go back to the original story. The quality department control over incoming customer specifications was put into operation. It soon became apparent that it was more successful than had been thought possible. The product engineers actually seemed to have been stimulated by its implementation. Following advice from the quality department, their actions improved considerably. It also became apparent, "through the usual channels", that the product engineers were very appreciative of the service they were receiving. As a general result, a first-class service of an unusual nature was being made available to customers.

When computer facilities at last became available, it was possible to begin planning the further check that had been waiting for years: to be able to search through thousands of customer specifications to determine if any might be affected by proposed process or product changes.

The procedure described in detail in Chapter 8 was developed and put into operation. There were a large number of customers, including all the well-known electronic equipment manufacturers. Once the system became operational, no case occurred in which an undesirable or unacceptable feature was missed during the quality department checks. On the other hand, it became known that other suppliers who had received the same specifications, and were supplying according to them, had missed many errors in them. It was clear that they were not operating any kind of control over incoming customer specifications.

There was an interesting and quite unexpected outcome of the operation of the new control system. In many instances it was obvious that it was causing at least some annoyance to customers. But in a rather perverse way, the company began to acquire an enhanced reputation with those customers, who reasoned that if this supplier was prepared to go to all that trouble and to risk upsetting them in so doing, the supplier must be eminently desirable compared to the others.

An indication of this attitude was provided by the number of times when after a query had been raised with a customer, its representative said: "It's funny you know. The so-and-so company is also supplying to the specification you queried, but they have not mentioned this point. I wonder if they have actually read our specification?" Such a reaction is a complete vindication of the application of such meticulous control over customer specifications quite apart from the twin goals of improving the quality of specifications and reducing LOST COSTS. It is worth the risk of strained relations, which are not, in any case, likely to remain that way for long. The relationship will improve as soon as the customer realizes what is actually happening.

Unfortunately, there are exceptions to every rule. Occasionally, one comes across a customer who refuses to consider any change to a specification, however justified it may be. In such a case, if the integrity of the supplier and his control are to be maintained, it will be necessary for the supplier to tell the customer that their order will not be accepted. (See current reports about major U.S. defense contractors being fined huge sums for similar "corrupt" practices).

The second example described briefly here happened before the system of control was introduced. It is indicative of the types of problems that companies can create for themselves as well as for their suppliers. A major equipment manufacturer had introduced a new system of specifications into one of their manufacturing divisions. It was intended to provide for components of three levels of reliability: high reliability , standard military reliability, and standard commercial reliability. Usually, this division of the manufacturer bought high-reliability components.

An order was received for the supply of components to specifications in the new series. The supplier was not absolutely sure which level of reliability was wanted, but as the customer usually bought the high-reliability type, that type was supplied. They were rejected as being not according to order. Therefore, it had to be the standard military class, and that was supplied. They, too were rejected as being not according to order. There was only one possibility left, the commercial class, and that was sent. That was rejected like the others.

It seemed that the customer's inspectors were not able to understand their own new specification system. There is a considerable moral here. This example is also another excellent case of LOST COSTS, costs that would not be shown in the company accounts under *any* heading.

# Case Study 1: Customer Specification Problems—A Solution

This chapter details the operational procedure that was developed to overcome the problem described in Chapter 7 concerning a large company handling many customer specifications. In a smaller company a much less complex solution would suffice. However, the author considers that this procedure provides a sound base from which other, simpler solutions can be developed for different sets of circumstances.

## A Procedure for the Control of Customer Specifications

1.  *Purpose.* This procedure has been prepared to ensure that:
    a.  There is a complete internal control over customer specifications.
    b.  They are so classified that unnecessary work is not expended upon them.
    c.  No feature in a customer specification that might be unacceptable to the company is accepted.
    d.  In the case of customer specifications which do not call up a company standard product, and which are accepted, company internal specification numbers are allocated and quality assurance acceptance specifications written.
    e.  The company quality assurance specifications so written are

prepared according to the standard format and rules contained in company procedure W.

f.  They are suitably coded and recorded in the computer memory system to facilitate the retrieval of all specifications in a particular group according to any selected code, or combination of codes, according to company procedure X.

g.  Customer specifications are properly looked after and filed according to an alphanumeric order by customer name.

**2.** *Scope.* All customer specifications are to be dealt with according to this procedure EXCEPT those which are clearly of draft status, have no identifying number, and are clearly intended for preliminary discussion purposes only. The company departments that are involved in the working of this procedure and upon which actions have been placed are as follows:

Central Marketing
Controller of Customer Specifications ($C^2S$)
The various Product Marketing groups
The various QA engineering groups

**3.** *Related Documents*

*Company procedure W:* rules and format for the preparation of company internal quality assurance specifications which are based on customer specifications
*Company procedure X:* system of coding of customer specifications, according to their specific requirements, for subsequent search and retrieval from the computer memory store
*Company procedure Y:* search and retrieval system, from the computer memory store, for customer specifications
*Company procedure Z:* computer recording system for customer specifications

**4.** *Operation of the Procedure*

**4.1.** Central marketing will ensure that all customer specifications that are received, except those mentioned as exceptions in paragraph 2 above, are sent to $C^2S$ and that an additional copy is sent to the relevant Product Marketing section.

**4.2** Upon receipt of customer specifications from Central Marketing, $C^2S$ will carry out the following actions:

**4.2.1** Examine all customer specifications and classify them as class B or class A according to the following rules, and so endorse them.

**4.2.1.1.** *Class B*

**a.** Specifications that call up company standard products, and which may or may not detail some or all of the mechanical or electrical parameters of those products.

**b.** Specifications that do not call up company standard products but whose requirements are simple and straightforward and deal only with mechanical or electrical parameters, and which may be selected out from a company standard product.

*Note:* Class B specifications may be "standard" or "nonstandard" with respect to company standard product.

**4.2.1.2.** *Class A*

**a.** Any general procurement specification

**b.** Specifications where there is a requirement for environmental testing

**c.** Specifications where there is a requirement for reliability testing

**d.** Specifications where there is a requirement for qualification approval

**e.** Specifications where there is a requirement for grouped testing

**f.** Specifications where there is a requirement for process change notification

**g.** Specifications where there is a requirement for special marking

*Note:* Class A specifications are automatically "nonstandard".

**4.2.2** C²S will deal with class B specifications as follows:

**4.2.2.1** *Standard Specifications*

**a.** Code them according to company procedure X.

**b.** Enter the details on the computer entry sheet, plus the note "accepted."

**c.** Enter details into the customer master drawing register (CMDR) plus the note "accepted."

**d.** File the specification.

*Note:* The minimum data entered into the computer entry sheet and the CMDR are:

Customer name
Specification number
Specification issue reference
"Accepted" or otherwise
"Standard" or "Nonstandard"
Company internal specification number, if allocated
Date of entry

**4.2.2.2.** *Class B "Nonstandard Specifications*

**a.** Code them according to company procedure X.

**b.** Enter the details on the computer entry sheet plus the note "pending."

**c.** Enter details into the CMDR plus the note "pending."

**d.** Enter the details of any nonstandard feature on the multipart Product Marketing Report Form (PMR).

**e.** Send copies 1 and 2 of the PMR to the relevant Product Marketing group.

**f.** Hold the specification as "pending" until the Product Marketing decision is made.

**4.2.2.3.** When class B "nonstandard" specifications are "accepted" by Product Marketing, $C^2S$ will take the following additional actions:

**a.** Receive copy 2 of the PMR back from Product Marketing with the "accept" decision and the allocated company internal specification number endorsed on it.

**b.** Delete the computer entry details under the note "pending."

**c.** Reenter details on the computer entry sheet, plus the note "accepted" and the newly allocated company internal specification number taken from copy 2 of the PMR.

**d.** In the CMDR, change "pending" to "accepted" and add the newly allocated company internal specification number taken from copy 2 of the PMR.

**e.** Enter the "accept" decision and the newly allocated company internal specification number on the retained copy 3 of the PMR.

**f.** Send copy 3 of the PMR to the relevant QA engineering manager to inform him that a new internal specification has to be prepared and to inform him of the details.

**g.** File the customer specification and copy 2 of the PMR.

**h.** Maintain a regular follow-up on Product Marketing until the company internal test specification has been written.

**i.** Destroy copies 4 and 5 of the PMR.

**4.2.2.4.** When the company internal test specification has been prepared by Product Marketing for a class B "nonstandard" customer specification, $C^2S$ will take the following additional actions:

**a.** Receive two copies of the company internal test specification from Product Marketing.

**b.** Check the test specification against the customer specification to verify that all the essential details have been included and that test limits for major parameters have been inset by the appropriate amount.

**c.** Pass one copy of the test specification and the relevant customer specification to the appropriate QA engineering manager so that he can have the QA acceptance specification prepared in accordance with company procedure W.

**d.** File the remaining copy of the test specification.

**e.** Maintain a regular follow-up on QA engineering until the QA acceptance specification has been written.

**4.2.2.5.** When the QA acceptance specification has been written by QA engineering, $C^2S$ will take the following additional actions:

**a.** Receive from QA engineering a draft copy of the QA acceptance specification and the customer specification.

**b.** Check the QA acceptance specification against company procedure W for correct format, and against the test and customer specifications for content.

**c.** Have the "master" copy of the QA acceptance specification prepared and obtain the necessary approval signatures (not initials).

**d.** Obtain and distribute copies of the QA acceptance specification.

**e.** File the "master" copy of the QA acceptance specification.

**f.** File the customer specification.

**4.2.2.6.** When class B "nonstandard" specifications are "rejected" by Product Marketing, $C^2S$ will take the following additional actions:

**a.** Receive copy 2 of the PMR back from Product Marketing with the "rejected" decision endorsed on it.

**b.** Delete the computer entry details under the note pending.

**c.** Reenter details on the computer entry sheet, plus the note "rejected."

**d.** In the CMDR, change "pending" to "rejected."

**e.** Advise the relevant QA engineering manager by sending him copy 2 of the PMR endorsed "rejected."

**f.** File the customer specification and copy 2 of the PMR. (Destroy all remaining copies of the PMR.)

**4.2.3.** C$^2$S will deal with class A specifications as follows:

  **4.2.3.1**

**a.** Code them according to company procedure X.

**b.** Enter the details into the computer entry sheet, plus the note "pending."

**c.** Enter the details into the CMDR, plus the note "pending."

*Note:* The minimum data to be entered into the computer entry sheet and the CMDR will be the same as those listed under para 4.2.2, except that these specifications will all be "nonstandard."

**d.** Enter into the PMR details of all nonstandard features, any other requirements thought to be undesirable or unacceptable to the company in their present form, and any statements considered to be factually in error.

**e.** Based on experience or knowledge, comment in the comment section of the PMR form as necessary on the details referred to in item d above.

**f.** Send copies 1 and 2 of the PMR to the relevant Product Marketing group.

**g.** Send copy 3 of the PMR to the relevant QA engineering manager for his information and then retain copies 4 and 5.

**h.** Hold the customer specifications as "pending."

**i.** Maintain a regular follow-up on Product Marketing until they have decided whether to "accept" or "reject" the specification.

*Note:* C$^2$S will ensure that he, and QA engineering, participate in any internal discussions with Product Marketing needed to determine a line of action when there are problems.

**4.2.3.2.** In the event that it is decided that external discussions with the customer are necessary to resolve problems raised in accordance with the actions in paragraph 4.2.3.1.d, C²S will take or coordinate the following actions:

**a.** Ensure that internal discussions have resulted in a common approach, by the various Product Marketing groups involved, which will be made to the customer.

**b.** Ensure that the customer is given advance copies of all the proposals to be put to him.

**c.** Coordinate all the arrangements for meetings with customer representatives.

**d.** Arrange for minute taking at any meetings with customer representatives and ensure that meetings move smoothly.

**e.** After meetings with customer representatives, ensure that all actions required of the company are taken.

**f.** After these meetings, maintain liaison with the customer, with suitable follow-up to ensure that the customer carries out all the actions required of, and agreed to by him.

**g.** As agreed amendments to the customer specifications are received, ensure that they are processed expeditiously, to ensure that there is speedy resolution of the problem.

**4.2.3.3.** When class A specifications are "accepted" by the Product Marketing group, C²S will take the following ad ditional actions:

**a.** Receive back from Product Marketing copy 2 of the PMR, with the "accept" decision and the newly allocated company internal specification number endorsed on it.

**b.** Delete the computer entry details under the note "pending."

**c.** Reenter details into the computer entry sheet plus the note "accepted" and add the number of the newly allocated company internal specification number taken from copy 2 of the PMR.

**d.** In the CMDR, change "pending" to "accepted" and add the newly allocated company internal specification number taken from copy 2 of the PMR.

**e.** Enter the "accept" decision and the number of the newly allocated company internal specification into the retained copies 4 and 5 of the PMR.

**f.** Send copy 4 of the PMR to the relevant QA engineering manager to inform him that a new internal test specification will be raised, and of the details.

**g.** File the customer specification and copy 2 of the PMR.

**h.** Maintain regular follow-up on Product Marketing until the internal test specification has been written.

**4.2.3.4.** When the company internal test specification has been written by Product Marketing, $C^2S$ will take the same additional actions as are called up in paragraphs 4.2.2.4.a to e for class B "accepted," "nonstandard" customer specifications.

**4.2.3.5.** When the QA acceptance specification has been written by QA engineering, $C^2S$ will take the same additional actions as are called up in paragraphs 4.2.2.5.a to f for class B "accepted" "nonstandard" customer specifications.

**4.3.** The relevant Product Marketing group for any particular customer specification is required to:

**a.** Consider the details and any comments passed to it by $C^2S$.

**b.** Participate in any necessary internal and/or external discussions.

**c.** Accept or reject the specification.

**d.** When an "accept" decision is made, a company internal test specification number must be allocated and an internal test specification must be written.

**4.4.** The relevant QA engineering group for any particular customer specification is required to:

**a.** Consider the detail and any comments passed on to it by $C^2S$.

**b.** Participate in any necessary internal and/or external discussions.

**c.** Prepare a QA acceptance specification when a company internal test specification has been written by Product Marketing.

# 9

# Case Study 2: Manufacturing Specification Problem

The company is a large one with some 3000 employees. It is in a high-technology industry manufacturing a number of classes of components on many production lines. A variety of components is produced on each line, and they are produced in both high and low volume and on both a batch and a continuous production basis. There can be as many as 60 process specifications on one production line. Often, some of the specifications will be used at more than one stage of the overall process on the line. Additionally, many of them are used on more than one line; thus, there is multiple usage.

A group of engineers is responsible for the techniques and processes on each line. Each individual engineer is responsible for all the processes on a part of the line. The actual running of the line is the responsibility of a production supervisor, who is responsible for its output. He is also responsible for ensuring that all the process specifications are followed. All the specifications relating to a particular line are issued in the form of a "book." The book is issued to all those who are considered by the manufacturing superintendent to have a need for it.

It was the duty of each individual book holder to ensure that his or her book is kept up to date as new specifications, and revisions to existing ones, are issued. Each book holder is required to ensure that only the latest issues of the specifications are used. Each book is subdivided into a number of sections, which are based on consecutive groups of operations generally common to all the production lines.

Much concern had been caused in the past by the outcome of audits which had been carried out on the general conditions of the specification books. In every case it had been found that there were problems. Specifications were missing; some were incomplete; some misfiled. Many explanations were put forth for this state of affairs, but none was very convincing. It was obvious that considerable cleanup was required, as the risks to the complex manufacturing processes were too great to allow the problem to continue.

The total number of specifications in the system was believed to be about 4500 and it was known that many were obsolete, although no one knew for sure, as there were no accurate records. They were classified according to a complicated nine-digit numbering system with a series of more than 40 vague descriptive classes into which a specification could be entered. It was possible to enter a specification into more than one class, depending on the interpretation of the engineer responsible for it. When a need for a new specification arose, it was almost impossible to determine if a suitable one already existed. It was also thought that there were many duplicates.

The existing system of control was clearly very ineffective and it was decided that a program to develop and introduce a new and effective system should be started, one that would be compatible with similar systems being introduced for other types of specifications. This program was expected to take many months for full implementation.

The first stage was to get rid of the deadwood. A list was compiled manually of every specification known to be in the system, listed by title and number. Copies of this list were given to each group of engineers with process responsibility. They were asked to strike from the list all specifications they were sure were no longer required in their processes. When the lists were returned they were collated carefully as, clearly, there would be many cases in which specifications not required by one line would be required on others. In parallel with this work, consideration was being given to the development of new numbering and titling systems.

It was decided to adopt a very much simpler system of classification based on the groups of processes in the subdivisions of the specification books. This reduced the number of classes from more than 40 to 9. At the same time, the numbering system was changed from nine digits to one with only five. The first digit identified the process group (from 1 to 9) and the remaining digits (from 0001 up) were to be allocated in consecutive order to specifications as they were issued, or reissued, in the new system.

As this work progressed many duplications were discovered. These resulted from the near impossibility of determining if a suitable specification already existed when a new one was required. In a few cases as many as four duplicates were found.

The new numbering system went only part of the way toward resolving the problem of the duplication of specifications. Consideration was then given to a new way of titling them, which could lend itself to alphabetical listing. If this should prove possible, it would be easy when a new specification was required to determine if a suitable one already existed. The existing titling method (although in truth it could not really be so described) was quite haphazard.

New titles were determined by the engineer who wrote the specification. More often than not the title chosen bore little relation to the purpose of the specification. Indeed, it was difficult to find any rationale behind the choice of title in many cases. No explanation was ever given for the oddities; perhaps the various engineers just liked the sound of the titles.

A large number of specifications whose titles did bear some relation to the contents was studied carefully. It appeared that a maximum of four words in each title was adequate to describe fully the purpose of the specification. Although there were often more than four words in the title, the extra words had little effect on the basic description of the purpose. An example that utilized four key words is the following:

Small signal planar transistor *wafer, first diffusion wash.*

The key words are in italic type. The actual "key" order was:

1. Wash
2. Wafer
3. Diffusion
4. First

It was decided that up to four key words would be ample to describe in full the purpose of a specification. Any other words used would be for aesthetic purposes and/or to placate the feelings of the engineers who were not overly anxious to lose the freedom of expression that had been theirs for so long. The key words would be underlined to identify them and would be the words entered into computer memory for that specification. In the same way, when the need arose for a new specification, up to four key words would be fed into the computer to determine if there were similar specifications already on file.

Several hundreds of the specifications whose titles had already been examined were now reexamined in the light of the new proposal. It was found that in almost every case it was possible to identify up to four words in each title that could be used to transform them as proposed. In most cases the titles did not need to be reworded. In some, including many not in the original sample, the titles had to be rewritten, as they did not describe accurately the purpose of the specification.

Limiting the number of key words to four, with the six digits in the new numbering system (the dash between the first digit and the following four counted as one character), made it easy to transfer the information into the computer. The use of computer facilities was a fundamental part of the process. Without them, and high-speed sort and printout facilities, provision of the alphabetical listing would not have been possible.

It was decided to provide a choice of two output facilities: either an alphabetical printout, or interrogation facilities via a visual display unit. To use the latter, one would feed in the key words of the proposed new specification. In response, the computer would list any or all existing titles, with numbers, which contained any or all of the key words. The questioner could then examine these specifications at leisure to decide if a new one really was needed.

Having disposed of the numbering and titling methods, one could begin to consider a control system for the specifications themselves. There would be fewer to control, as removal from the system of the obsolete and duplicate specifications had reduced the numbers by about 20%.

The operating procedure developed for control purposes is detailed in full in the next chapter. It is the actual procedure that was put into use: none of its apparent complexities were "dreamed up" for the purpose. One of the apparent complexities is that no fewer than 10 individual task functions, or functional groups, have specific responsibilities under the procedure. They were all required.

# 10

# Case Study 2: Manufacturing Specification Problem—A Solution

## A Control Procedure for Manufacturing Specifications

**1.** *Purpose.* The purpose is to define the contents of; the methods used for the control of; the preparation, issue, and distribution of; the manufacturing/process specifications for; and the functions of Specification Services (SS).

**2.** *Scope.* Within the field bounded by the following definitions for manufacturing specifications and the SS, this procedure concerns the production, purchasing, and quality departments.

**3.** *Policy.* For any of the company's products that are in full production, excluding pilot line or experimental stages, the manufacturing processes will be supported in detail by approved manufacturing/process specifications.

**4.** *Definitions*

    **4.1.** *Manufacturing/Process Specifications.* Each specification contains all necessary information about, and describes any one stage, or all stages, of a process or assembly or test operation that must be carried out in terms of the overall manufacturing process of a company product. It can also be a drawing or other written material or detail of parts or material either incorporated in the product or required as part of the manufacturing process. Except in the case of piece parts and other materials, the specifications will use the format and contain the information indicated below:

    **a.**  Equipment required, with drawing numbers

    **b.**  Materials required, with part numbers

    **c.**  Equipment setup

    **d.**  Procedure to be used

    **e.**  Details of process control and/or surveillance

    **f.**  Inspection requirements

**4.2.** *Specification Services (SS).* This is a group that has responsibility, as defined later, for manufacturing specifications only.

**4.3.** *Flow Diagram.* Flow diagrams are to be used, with standard ASME symbols, and will list in sequential order the various stages in the manufacturing operation, the manufacturing specifications relating to the various stages and details of the parts and materials that are used to make the product. It is, in itself, considered to be a manufacturing specification.

**4.4.** *Product Specification.* This is a document that contains the mechanical and electrical characteristics of the finished product and should correspond, approximately, to the product data sheet for technical content.

**4.5.** *Parts Drawing.* These drawings give full details of piece parts required for the manufacture of the product.

**4.6.** *Product Drawing.* These drawings illustrate the assembly buildup and the interrelationship of individual parts at any given stage of the process.

**4.7.** *Engineering Change Notice (ECN).* A document used to initiate new manufacturing specifications and changes to existing ones. Requires a variety of supporting data relating to urgency, piece-part obsolescence, and the like, plus an "action" date. Is a four-part document in numbered sets in pads.

**4.8.** *Temporary Engineering Change Notice (TECN).* A document similar to that just described except that it is used to introduce temporary changes only to a process and has a maximum life of 30 days. If starting and terminating dates are not given, it has an automatic life of 7 days only. It is a three-part document in numbered snap-out format.

**4.9.** *"Strictly Private" Classification.* This classification will be applied to all specifications for processes and materials considered to be technically ahead of the rest of industry. These must be closely guarded to prevent disclosure of the contents to unauthorized persons. They may be sent outside the company only on the written authority of the managing director.

**4.10.** *"Internal Data" Security Classification.* This classification is used for specifications and material generally known throughout the industry. They may be sent outside the company on the approval of the relevant engineering manager.

**5.** *Individuals with Responsibilities Under the Procedure*
- **5.1.** Managing Director
- **5.2.** Product Department managers
- **5.3.** Engineering managers
- **5.4.** QA engineering managers
- **5.5.** Manufacturing managers
- **5.6.** Product engineers
- **5.7.** QA engineers
- **5.8.** Production planners
- **5.9.** Individual specification book holders
- **5.10.** Specification Services (SS) supervisor

**6.** *Responsibilities of Individuals*

**6.1.** *Managing Director.* As required, authorizes "strictly private" specifications to be taken or distributed outside the company.

**6.2.** *Product Department Managers.* Approves and updates "responsibility" lists.

**6.3.** *Product Engineering Manager*

**6.3.1.** Approves by signing (no initials) all engineering change notices (ECNs) and new and changed manufacturing specifications. May also delegate this authority for a specific period. The name of the person to whom authority is delegated, and the dates of the specific period, must be given in advance to the SS supervisor in each case.

**6.3.2.** Determines the security classification of individual manufacturing specifications.

**6.3.3.** If the manager changes the security classification of an individual specification, issues an ECN to the SS supervisor so that the change can be actioned.

**6.3.4.** As required, authorizes the lowest level of security classification of specification, "internal data," to be taken or distributed outside the factory.

**6.3.5.** Within the basic framework of the system, determines how manufacturing specification books shall be sub-divided.

**6.3.6.** Decides on the initial assignment of manufacturing specification books (bearing in mind the limitation in numbers of copies permitted; see paragraph 9.1) and any excess distribution. Determines the general basis for the distribution of manufacturing specifications and authorizes new or reassigned engineers to receive copies of specific manufacturing specifications.

**6.3.7.** Authorizes the cancellation of a manufacturing specification book if it is no longer required.

**6.4.** *QA Engineering Manager*

**6.4.1.** Ensures that QA engineering assistance is provided to product engineers whenever process control, surveillance, or other inspection activities are incorporated in a manufacturing specification.

**6.4.2.** Approves by signing (no initials) all manufacturing specifications that incorporate process control, surveillance, or other inspection activities. May also delegate this authority for a specific period. The name of the person to whom authority is delegated, and the dates of the specific period, must be given in advance to the SS supervisor in each case.

**6.5.** *Manufacturing Managers*

**6.5.1.** Assigns responsibility to qualified personnel to ensure the use of latest issues (only) of manufacturing specifications.

**6.5.2.** Ensures that the line supervisors sign all ECNs relating to their lines as verification that they have seen the details of each ECN and are aware of its effect on them.

**6.6.** *Product Engineers*

**6.6.1.** Must determine if a manufacturing specification is required. If it is, must determine if a suitable one already exists. If not, will write a new one.

**6.6.2.** Will write flow diagrams for the complete process using standard ASME symbols.

**6.6.3.** Writes an ECN introducing each new specification to the relevant flow diagram.

**6.6.4.** Ensures that all manufacturing specifications are referred to the flow diagram(s) on which they are called up and to the manufacturing specification books and subdivisions thereof in which they appear.

**6.6.5.**   When changes are to be made to any stage of a process, will write permanent or temporary (as the case may be) ECNs. In the case of a temporary ECN, a time limit not exceeding 30 days for its use must be set.

**6.6.6.**   Ensures that all interested parties are made aware of the fact that ECNs become effective immediately upon being signed.

**6.6.7.**   Ensures that all ECNs receive a security classification and then receive all necessary signatures (no initials) of authorization, and that any change of class is indicated.

**6.6.8.**   Sends the top copy of all authorized ECNs to the SS supervisor for typing, allocation of number, and subsequent issue. Also sends a copy with all relevant information to production control, who are required to check material stocks and program.

**6.6.9.**   When new or amended specification masters are received from the SS for checking and *signing off*, the engineer must obtain the necessary signatures of authorization (no initials) and return the masters to the SS within three working days.

**6.6.10.**   Ensures that the content of manufacturing specifications and flow diagrams is kept under constant review and up to date.

**6.6.11.**   Ensures that as specifications cease to be used, an ECN is drawn up withdrawing them from use so that the SS supervisor can remove them from the system.

**6.6.12.**   Prepares product specifications as necessary.

**6.6.13.**   Prepares product drawings as necessary.

**6.6.14.**   Prepares piece-part drawings as necessary.

**6.6.15.**   Must ensure that the stock, cost, and availability position of piece parts and materials is always taken into account when process changes are being considered.

**6.7.**   *QA Engineers*

**6.7.1.**   In each case when an ECN concerns process control, surveillance, inspection, or any other aspect of Q.A., will be required to sign the ECN to indicate the approval of QA.

**6.7.2.**   In the case of all permanent ECNs, will examine them to determine whether, and what, there is in them relevant to the data retrieval system of the customer drawing register, and then suitably advise the CDR supervisor.

**6.7.3.**   Will also examine permanent ECNs to determine

whether the change proposed, or introduced, is one that will require requalification testing to a military or customer specification. If so, must draw the attention of product engineering, marketing, and the QA qualification approval section to the need for such requalification so that it can be undertaken.

**6.8.** *Production Planners*

**6.8.1.** Advises Purchasing of expected piece-part changes as far in advance as possible. Must also advise all interested personnel of any changes in the rate of usage of piece parts or materials.

**6.9.** *Individual Book Holders*

**6.9.1.** Acknowleges receipt of copy, or copies, of a manufacturing specification book or subsection thereof.

**6.9.2.** Ensures the security of any book(s) allocated to him or her.

**6.9.3.** Must not take book(s) out of the plant.

**6.9.4.** Returns book(s) to SS if transferred out of the section.

**6.9.5.** Ensures that any additional copies of specifications obtained are not used after a later issue of the specification has been issued.

**6.9.6.** Ensures that specifications are filed in his or her book(s) in numerical order, with the flow diagrams coming first.

**6.9.7.** Ensures that all advance copies of ECNs issued to him or her are filed with the relevant specification until a copy of the revised specification is received.

**6.9.8.** Ensures that all copies of earlier issues of specifications in his or her book(s) are removed and destroyed as new issues are received.

**6.9.9.** Consistent with security requirements, ensures that his or her books are reasonably accessible to all who have a legitimate need to use them.

**6.10.** *Specifications Services (SS) Supervisor*

**6.10.1.** Date-stamps ECNs as received from the product engineers.

**6.10.2.** Examines ECNs to verify compliance with respect to approval authorizations, change class, and the like, plus security classification.

**6.10.3.** Reads ECNs to ensure that they can be copy-typed "straight off" without queries. Any ECNs with queries will be returned to the originator for clarification.

**6.10.4.**  Will verify that all specifications introduced by an ECN are (a) already included in flow diagram(s) or (b) accompanied by a separate ECN adding the new specification to the relevant flow diagram(s). Any specification not referenced to a flow diagram will be returned to the originator with its ECN for the appropriate action to be taken.

**6.10.5.**  Will also verify that all specifications are referenced to manufacturing specification book sections. If not, they will be returned to the originator for appropriate referencing.

**6.10.6.**  Alters the layout of a specification as necessary to ensure reasonable conformity of format with others of the same type.

**6.10.7.**  Allocates a reference number to each new manufacturing specification according to the coded numbering system.

**6.10.8.**  Enters details of the specification title against its reference number in the manufacturing specification register.

**6.10.9.**  Has new specifications, and changes to existing ones, typed from the ECNs and sends to originator for checking and to obtain all necessary approval signatures (not initials).

**6.10.10.**  Arranges for flow diagram(s) to be drawn, using standard ASME symbols, and any other drawings or diagrams that need to be prepared from engineers' sketches on the ECN.

**6.10.11.**  Initiates "chasing" action if typed master copies of specifications are not returned by the originator with all necessary approval signatures (no initials) within three working days.

**6.10.12.**  Files all "actioned" copies of ECNs for a period of 12 months. After this period on file, "actioned" copies of ECNs will be destroyed.

**6.10.13.**  In consultation with the engineering managers will determine:

**a.**  The breakdown of the manufacturing specification books into subsections and the composition of these new subsections.

**b.**  The distribution for complete manufacturing specification books, including their total number.

**6.10.14.** Obtains book holders' signatures on a receipt form, which also signifies his acceptance of the responsibilities of a book holder.

**6.10.15.** After new and changed specification master copies have been approved, obtains the required number of copies and distributes them via the company's internal mail system to the authorized specification book holders indicated.

**6.10.16.** Files all master copies of manufacturing specifications by numerical order of their reference numbers.

**6.10.17.** Carries out audits on manufacturing specification books at suitable intervals to determine correctness of contents. Engineering managers and other relevant personnel will be informed of the results of audits.

**6.10.18.** At suitable intervals removes obsolete or redundant specifications from the files, microfilms them, and destroys the originals. The microfilm will be filed suitably and stored to permit reasonable access. Will make a survey from time to time to ensure that product engineers are declaring unwanted or unused specifications obsolete.

**6.10.19.** Prepares and circulates a "contents" list for each manufacturing specification book, or section thereof, based on the relevant flow diagrams. This contents listing will be on a numerical basis and will be updated periodically.

**6.10.20.** As the occasion or need arises, provides any other information breakdown, retrieval, or display system for manufacturing specifications.

**6.10.21.** As the occasion arises, and to cater to future developments, provides a microfilm library of all current specifications.

**6.10.22.** Provides the purchasing department directly with copies of any revised drawings for purchased parts, or for any other specification that concerns them or appears to concern them.

**7.** *Documents That Are Required to Be in Manufacturing Specification Books*
    **7.1.** Flow diagrams
    **7.2.** Manufacturing/process specifications
    **7.3.** Piece-part drawings
**8.** *Documents That May Be in Manufacturing Specification Books*
    **8.1.** Product specifications

**8.2.** Product drawings
**8.3.** Engineering change notices
**8.4.** Temporary engineering change notices
**9.** *Permitted Number of Manufacturing Specification Books*
**9.1.** *Product Department.* A maximum of four complete books for each product or line, increasing to five books when the QA function (process surveillance) is operating in the production department.
**9.2.** *Quality Assurance Department.* A maximum of two complete books for each product or line, decreasing to one when the QA function (process surveillance) is operating in the production department.
**9.3.** *Subsections of Manufacturing Specification Books.* Each book is divided into a number of subsections. These are not counted separately when determining the number of books. The number of books referred to in paragraphs 9.1 and 9.2 are maxima; whenever possible, fewer are to be prepared and issued.
**10.** *Subdivision of Manufacturing Specification Books*
**10.1.** As, in most cases, no person needs a complete set of specifications for a product or line, the books are sectionalized on a basis that fits the manufacturing process, as follows:

*Section 1;*  Material preparation
*Section 2;*  Material processing (1)
*Section 3;*  Material processing (2)
*Section 4;*  Material processing (3)
*Section 5;*  Material processing (4)
*Section 6;*  Assembly
*Section 7;*  Final test
*Section 8;*  Purchased parts inspection
*Section 9;*  Plating

**10.2.** At the discretion of the engineering manager concerned, two or more sections may be combined when, for example, all material processing, sections 1 to 5 inclusive, is carried out on a single-line-responsibility basis.
**10.3.** The engineering managers may also, at their discretion, subdivide the books on a product, instead of a process, basis.
**11.** *General Information*
**11.1.** To the maximum extent possible, engineers should use

existing specifications rather than write a new one merely because there appears to be a need for one for a new product.

**11.2.** Manufacturing specifications should always be supported by "used on" information.

**11.3.** No updating service is provided for casual copies of manufacturing specifications, and it is the responsibility of the recipient to ensure those extra copies are not used in conflict with the general distribution procedure.

# Case Study 3: In-House Acceptance Specification Problems

An audit had been made into the handling of the acceptance specifications in the final acceptance area of a product manufacturing group. It was the first such audit that had been carried out for a considerable time. It had been intended primarily to test the general conditions with regard to the use and storage of the acceptance specifications. There had been a feeling that the acceptance supervisor had been rather lax in this respect. As it happened, the picture that began to appear as the audit progressed prompted an extension to its scope well beyond the original intention.

The following items were now included: specification preparation, including format and revision procedures; and the preparation, issue, and revision of the program tapes for the computer-controlled automatic test equipment. Not all of the components were tested on the automatic computer-controlled equipment, but a substantial proportion was.

Beginning with the original aim of the audit, it was found that, occasionally, the acceptance test record cards were prepared without reference to the appropriate product acceptance specification. Insufficient care was taken with the filing and storage of the specifications. They were serially numbered, but misfiling was frequent, and at times, only some pages of a specification were filed. Some of the specification copies had gone beyond what could have been called their useful life, so that their use was difficult. No attempt had been made to get new copies. In this limited area of the audit, alone, the result was clearly unsatisfactory.

However, that result, as it turned out, was much less disturbing than other facts that appeared. There was no formal procedure for the preparation of the acceptance specifications, and there were many differences in format and presentation. These differences did not make for their easy use. It was obvious that specification preparation was a task that was shared by a number of engineers, dependent on individual availability. Nor could one even depend on the same engineer using the same format for successive specifications. Much seemed to hinge on their state of mind at the time they began to prepare a new one.

In a few cases, checks back to the original customer specification showed that there were sometimes omissions. Sometimes, too, the customer's requirements were not interpreted correctly. Much more serious, however, was the number of unauthorized, unsigned, and undated amendments that had been made to many of the specifications. Some were made in ink and some in pencil, and there did not seem to be any formal record of any of the changes. Clearly, there was a considerable lack of control.

There was supposed to be a procedure for completing and filing the acceptance records for each type of component. Yet in a small sample of acceptance batches chosen from those known to have been accepted recently, few of the record cards could be found. Hasty searches turned up about two-thirds of the missing cards. Some had not been filed after many weeks. Some had been misfiled. The remaining third could not be found after extensive searching. It was also noted that occasional errors had occurred in the transfer of information—batch identification numbers—from one set of records to another. This was serious from the point of view of batch traceability in the event of product faults arising.

Of all the faults found, those relating to the test tapes were by far the most serious. Preparation of the programs from which the tapes were prepared was a skilled job and one that required careful checking. Once a tape had been prepared, it was a time-consuming job to get a printout for the purpose of checking against the original program. The language used was incomprehensible to laypersons. Recognizing those difficulties and the cost consequences arising from the use of an incorrect tape, there was a procedure for program and tape testing, although it was rather informal. The result of the audit made it clear that this informal procedure had broken down completely.

In a number of cases, test tapes had been duplicated—in one case, triplicated—by tapes bearing the identical references of the "true" tapes but with different tests and limits programmed. There was no reasonable way for the QA acceptance personnel who operated the automatic test equipment to verify any of the tapes if they found two with identical references.

In the explanation for extra tapes, the lack of thought displayed by the engineers responsible was almost unbelievable. They had been prepared for engineering evaluation purposes on product samples produced to experimental specifications. The tests and limits on them were very different from the acceptance tests and limits on the "regular" tapes. Even more unbelievable was the fact that these experimental tapes had been given the same identification references as the correct tapes. When they were finished with them, the engineers, who ought to have destroyed them or at the very least kept them under lock and key, actually left them with the acceptance tapes. The acceptance personnel did not even know that these special tapes had been prepared and left with their regular tapes. They did not know that there was any difference in the identical tapes which they sometimes found that they had. They just assumed that they were duplicates, a pair of identical tapes that were used interchangeably.

Not surprisingly, a number of corrective actions were instituted when the full results of the audit became known. These included new and strict control procedures for the preparation of in-house acceptance specifications from customers' specifications, and even stricter controls for the preparation of the test tapes. The latter was very carefully designed to ensure that it was virtually impossible for an unauthorized tape to be used for acceptance purposes. Improvements were also made in the relatively minor systems for the use and storage of the acceptance specifications in the final acceptance area, for the completion and filing of the acceptance records, and for the transfer of information from one set of records to another when required.

Apart from the discovery of the various system faults that were found as a result of the audit, there was one other important outcome: demonstration of the real value of an audit procedure for determining if things are being done as they should be—in this case, the preparation and use of specifications.

In the next chapter, the full company procedure is detailed which was developed for controlling the preparation of in-house acceptance specifications from customers' specifications. Again it may be noted that although this procedure is tailored to suit a fairly complex situation in a large supplier company dealing with very many customers, there should be no problem in using it as a base for simpler systems for simpler problems and smaller companies.

# Case Study 3: In-House Acceptance Specification Problems—A Solution

## A Control Procedure for In-House Acceptance Specifications

**1.** *Purpose.* The purpose is to specify a standard format for the presentation of quality assurance X final acceptance specifications, the type of information that may be included in them, and guidance about the parts of the X specification into which the various kinds of information are to be included. The intent is to make it easier for those who write the specifications and, more importantly, easier for those who will have to read, and work from, them.

**2.** *Scope.* The procedure applies only to the engineering and final acceptance branches of the quality assurance department.

**3.** *Operation of the Procedure.* The X specifications will be prepared in 11 sections. The purpose of these sections, together with details of the information that may be included in each, will be detailed in later paragraphs of this procedure. The order of the sections has been selected to ensure, as far as possible, that the presentation of the information will best suit the needs of its users in the QA department. The X specifications *must be written directly from* the customers' specifications.

**3.1.** *Section 1: Customer Information.* The name of the customer and the number of his specification must be given. Care should be taken to ensure that, in those cases to which it applies, subsidiary specifications that may be called up on the customers' main specification are also listed. In those cases the titles of the subsidiary specifications must also be given. This information is helpful to the reader and provides a guide to the relative importance of the various specifications. As some customers' specification numbers are made up of complex groupings of alphanumeric characters, care must be taken to ensure that they are fully and accurately detailed.

**3.2.** *Section 2: Basic Component Type.* It is usually the case that a customers' requirement is met by selection from an existing standard component. When this is so, the type number of the standard component is to be given in full. Occasionally, the customer's requirement will be met from an existing X, or other special selection, specification. In those instances the number of the relevant specification is to be given. In all cases an indication is to be given as to whether electrical testing and/or mechanical work is required. (Mechanical work in this context is to include any special marking, which may be the only requirement of the customer.)

**3.3.** *Section 3: Customer Special Requirements.* Customer specifications occasionally call up requirements, or request the use of procedures, which the company does not normally provide. Examples of these are the following:

**3.3.1.** Notification of process changes. (It should be noted that ordinarily, the only process changes about which the company will agree to notify the customer are those which, if they were made to a military specification component, would require that the component would have to be requalified.)

**3.3.2.** Special markings on primary containers, such as plastic envelopes.

**3.3.3.** Special documentation. In these cases references must be made in this section to the fact that these special requirements exist. The detail regarding special markings and documentation is given in sections 4 and 7, respectively.

**3.4.** *Section 4: Identification and Special Markings Requirements.* Invariably, a component that is to be selected out for any requirement from any standard or other special selection component must be identified in such a way that it cannot be confused with any other component. When the customer does not specify special marking, the component should be identified by the X specification number only. This number may be qualified, however, by suitable letters or figures to indicate the application of any of the package options for the component which are available. (These include transit carriers.)

When the customer does specify markings that he wants to have on the component itself or on any of the packaging, the exact details must be given, including any special requirement for the location of the marking. The customer's requirements should be studied carefully to determine if they might prevent the company from applying its own standard identification marking features, which are applied whenever physically possible (i.e., the company trade mark and the date code). Occasionally, a customer's requirement will exclude any company marking. Sketches may be used to clarify the customer's requirements but only if it is absolutely essential.

**3.5.**  *Section 5: Component Configuration and Package Details.* This section should contain all the necessary information about the package in which the component is to be supplied and about whichever of the options mentioned in the preceding section have been selected. As far as possible, drawings and sketches should not be used and references to packages should be to British or other standard requirements. Other than this, only written descriptions should be given of the various options. Sketches may be used only in those rare cases when the customer has ordered a nonstandard type of packaging which the company has agreed to supply.

**3.6.**  *Section 6: Relevant, Associated, Company Specifications.* It is often the case that time and effort are saved in writing specifications by reference to other specifications, rather than including possibly lengthy extracts from them. These specifications should be referenced by number and title (e.g., "10-505. Final acceptance, visual and mechanical standards. Section 11; para 3.11.3.2."). It must also be pointed out that when other specifications are quoted, care must be taken to ensure that they do apply and that only the relevant sections are quoted.

**3.7.**  *Section 7: Special Records.*   Reference was made in section 3 to special documentation requirements that might be specifically demanded by a customer. Details are given in this section. There are many different types of documentation that a customer may demand. Examples are given in the following subparagraphs:

**3.7.1.** The components are to be individually tested, actual results recorded, and a copy of the test results sent to the customer before (or after) shipment or with the shipment.

**3.7.2.** Approval by the customer of the suitability of components is to be confirmed in writing before shipment is made.

**3.7.3.** A requalification exercise of some type or other has to be undertaken at specified intervals of time or frequency of shipment.

**3.7.4.** A processing sequence may be specified that is different from standard. It may also specify QA clearance at certain stages in the sequence.

**3.8.** *Section 8: Flow Diagram.* It is desirable, especially when there is a customer requirement such as that in paragraph 3.7.4, that there be a listing of the sequence of the manufacturing operations, with identification of key QA and/or government quality department checkpoints. When this is so, a flow diagram should be prepared showing the operational sequence and the location of any mandatory checkpoints. The operational descriptions should be full and include the specification numbers where these are relevant. Standard ASME flow diagram symbols should be used.

**3.9.** *Section 9: Test Equipment.* Only major items of equipment that have to be used should be mentioned here: for example, automatic test equipment, or environmental facilities. Minor items of equipment are not to be listed.

**3.10.** *Section 10: Tape and/or Program Details.* In the case of many components that are produced in large volume, the final testing is carried out on automatic computer-controlled test equipment. There may also be more than one program for a component to cover; for example, testing at different temperatures. It may also be the case that certain types of test cannot be carried out on automatic equipment. Therefore, the full identification of any required program must be given together with its function (e.g., high-temperature testing). Any groups of tests that are not performed on the automatic test equipment must also be indicated. In any instance in which there are no test programs, this fact must be indicated by writing "No test programs. All tests conducted manually."

**3.11.** *Section 11: Q.A. Acceptance Tests.*

**3.11.1.** In a few cases the reason for raising an X specification is because of mechanical differences. Electrical testing is exactly the same as for an existing standard component or for one or another X specification. In those cases the details of the testing are not given. All that is said is: "Electrical testing exactly as for. . . . "

**3.11.2.** If there are only one or two differences in testing from an existing specification, a statement such as "electrical testing as for . . . except for the following tests" will suffice. "The following tests" must be specified. In these two instances the information in sections 9 and 10 should reflect the position.

**3.11.3.** In the majority of cases, electrical testing will be significantly different from that required for any existing component, and full details will have to be given. The method to be used is described in the following subparagraphs:

**3.11.3.1.** A system of subgrouping of the tests is to be used which is broadly based on that used in military specifications for the same types of components, although not necessarily identical. Four main groups of tests will be called up: groups A and B, and occasionally C and D. Each will be subdivided into a number of subgroups. These subgroups will be fixed as far as contents are concerned for those subsidiary to A and B. Subgroups of C and D will be variable in content. Group A will be used for normal final acceptance and consist primarily of electrical testing, group B for environmental types of tests, group C for short-term or periodic testing, and group D for long-term or infrequent periodic testing.

**3.11.3.2.** *Group A.* There will be a maximum of seven subgroups, although they will not all necessarily be used at the same time. Each will have a particular function and they are *not* interchangeable. This means that each must be referred to even if only to insert "no tests required" against it.

Subgroup 1 is for visual and mechanical inspection, including dimensions. To avoid the use of unnecessary detail, the relevant QA specification may be called up (e.g., "10-505. Final Acceptance, Visual and Mechanical Standards"). It will be necessary to be sure that it does include dimensional information, or it will be necessary to refer to the relevant BSI drawing.

Subgroup 2 is limited to functional testing only and to the minimal amount of testing which will verify that the component actually functions. Ordinarily, it will not be used when full final acceptance is to be carried out. Its use should be restricted to those occasions when components that have been fully "final accepted" in one configuration are being reworked to change the configuration (e.g., the component lead-outs are being formed into a nonstandard shape). Thus this subgroup contains only these electrical tests that are needed for reacceptance. In these cases subgroup 1 will also apply.

Subgroups 3, 4, and 5 are broadly similar in that they all call up dc parametric tests but at different temperatures. Subgroup 3 is for low- temperature tests, 4 for ambient temperature testing, and 5 for testing at elevated temperatures.

Subgroup 6 is for ac and switching tests which have to be undertaken on an individual test basis on nonautomatic test sets.

Subgroup 7 is provided for any other electrical testing that does not fall into any of subgroups 2 to 6 inclusive and would not ordinarily contain any test requirements.

**3.11.3.3.**   *Group B.*   In this group, too, there will be a maximum of seven subgroups which will not necessarily all be used in any one specification. As for the subgroups in group A, each will have a particular use, will not be interchangeable, and will be referred to in every specification, if only to insert "no tests required" against it.

Subgroup 1 is for vibration fatigue.

Subgroup 2 is for constant acceleration.

Subgroup 3, is for lead fatigue.

Subgroup 4 is for shock.

Subgroup 5 is for life testing, operational, at low, ambient, and elevated temperatures.

Subgroup 6 is for nonoperational life testing, as for subgroup 5.

Subgroup 7 is for any other environmental test that may be called up, which is not in any of subgroups 1 to 6 inclusive. Ordinarily it would be endorsed as having "no test requirement."

Many of the tests in group B have posttest endpoints. They may be given individually in each subgroup as required or after a "group" of subgroups if the same posttest endpoints apply to more than one subgroup.

**3.11.3.4.**   *Groups C and D.*   On the infrequent occasions on which these groups are required, the subgrouping of tests will be as in paragraph 3.11.3.3 as far as possible. Contrary to the practice for unused subgroups of groups A and B, unused subgroups of C and D will not be listed.

Acceptable quality levels (AQLs) and inspection levels will normally apply to a subgroup collectively. For each group they will be tabulated before subgroup 1 in each case. In the group A subgroups, the various tests will be listed on a line-by-line basis, giving, as required, the following information in each case:

**a.**   Test name
**b.**   Test symbol
**c.**   Test conditions
**d.**   Test limits

In the group B subgroups the same procedure will be followed except that the test conditions may be referenced to the appropriate paragraphs in the controlling military specification for the class of component. In the case of groups C and D, they will be dealt with as seems appropriate within the rules for groups A and B.

# 13

# Case Study 4: Control Procedure for Procurement Specifications

The occasion for the preparation of this control procedure was not that a critical situation had arisen, as was the case with the earlier case studies. Rather, there was a desire to extend the series of controls as widely as possible over the various specification systems. The hope was that this would prevent future problems. In developing this particular procedure, the experience gained in the development of the earlier procedures, described in Chapters 8, 10, and 12, was taken into account and was valuable.

## A Control Procedure for Procurement Specifications

**1.** *Purpose.* This procedure has been prepared to ensure that control is exercised over the preparation of procurement specifications for the company by procurement specification control (PSC). By the exercise of this control it is expected that the company procurement specifications will benefit to the extent indicated in the subparagraphs that appear below—and more importantly, that the company's technological image will be enhanced for both suppliers and customers. It is also anticipated that as a result of the operation of this control, there will be a significant reduction in LOST COSTS. These are the costs which would ordinarily be incurred in making unnecessary amendments to procurement specifications to match the capabilities and inclinations of suppliers.

**1.1.** *Benefits That Are Expected to Accrue from Use of the Procedure.*

**a.** The task of users will be eased by use of a standard format

**b.** Maximum simplicity of presentation.

**c.** The use of standard language which will be clear and unambiguous and suitable for the type of user for whom it is intended.

**d.** The requirements placed on suppliers will correspond to their capabilities.

**e.** The use of logical and coherent titling, numbering, issue, and revision systems for the specifications will simplify the process.

**f.** All departments in the company that might have an interest in any particular specification will have had an opportunity to comment on it before it is finalized and issued.

**2.** *Scope.* The departments of the company that are normally directly interested in the preparation of, and usually produce, first drafts of procurement specifications are:

Design, Product Engineering, Production

Other departments that will, or may, have an interest in their preparation will include as a minimum the following:

Quality Assurance, Purchasing

From time to time, other departments may be considered to have an interest and will be consulted as required.

**3.** *Related Documents*

**3.1.** *Company procedure a:* rules and format for the preparation of company procurement specifications for electronic components

**3.2.** *Company procedure b:* procedure for titling, numbering, issue, and revision of company procurement specifications for electronic components

**4.** *Operation of the Procedure*

**4.1.**

**a.** PSC will receive drafts of proposed procurement specifications from one of the originating departments mentioned in paragraph 2. Exceptionally, drafts may originate in other departments, such as QA.

**b.** PSC will first classify the drafts so that each may be placed in one of the three classes C, B, or A by applying the following rules:

**i.** *Class C.* Specifications in this class will be those that are simple and straightforward and specify no more than normal mechanical and electrical parametric requirements. These components are those suitable for use in normal commercial equipment.

**ii.** *Class B.* These are specifications that call up environmental tests and/or complex electrical tests in addition to the tests of a class C specification. This class of component will be that used in professional and ground-based military equipment.

**iii.** *Class A.* In addition to the requirements for a class B specification, a class A specification will call up high reliability requirements, qualification approval, process change notification, recorded test results, parametric change data, and similar items. The components will be used in critical and airborne military equipment, aerospace equipment, and industrial equipment with very high reliability requirements (e.g., telephone service underseas repeaters).

**4.2.** PSC further actions will depend on the class into which the specification is placed and will be in accord with the following paragraphs.

**4.2.1.** *Class C*

**a.** Rewrite the specification in standard format.

**b.** Send copies to the QA and purchasing departments for comment. (Purchasing may wish to add notes about recommended suppliers.)

**c.** Subject to any comment from the QA and/or the purchasing departments, a master copy of the specification is prepared, number allocated, and title determined and it is sent to the originator to be endorsed with the necessary approval signatures (no initials).

**d.** After approval of the master copy, copies will be taken and sent to the originator as a record and filed by PSC, as also is the master copy.

**e.** The QA and purchasing departments are informed that the specification is now available on requisition.

**4.2.2.** *Class B*

**a.** Check any environmental test requirement to verify that it conforms to one of the recognized tests in BS 2011, MIL-STD-883, or other standard, and ensure that they are identified as such. If the stipulated tests are different or more severe, the originator will be required to justify the departure from the use of standard tests. (Their use will result in additional cost.)

**b.** If any complex electrical tests are required, check whether prospective suppliers will have suitable equipment to perform the tests.

**c.** Rewrite the specification in standard format.

**d.** Send copies to the QA and purchasing departments for comment. If nonstandard tests are required, Purchasing would be expected to comment on possible supply difficulties. QA and Purchasing are both expected to comment on any special quality or reliability features (such as data logging), with particular reference to any problems that these tests might be expected to give rise to with suppliers. (There is no point calling up special tests if it is probable that there is no supplier that can carry them out.)

**e.** In the event that comments are received which have to be resolved, PSC will coordinate internal company discussion to remove the difference(s).

**f.** If need be, and with advice from QA or Purchasing or both, PSC will initiate discussions between company representatives and probable suppliers to ensure a final version of the specification which, while meeting the requirements of the originator, will also be acceptable to suppliers in its final form.

**g.** A master copy of the specification will be prepared, number allocated and title determined, and sent to the originator for endorsement with the necessary approval signatures (no initials).

**h.** After approval of the master copy, copies will be taken and sent to the originator as a record and filed by PSC, as also is the master copy.

**i.** The QA and purchasing departments are informed that the specification is now available on requisition.

**4.2.3.** *Class A.* The various steps that are taken to deal with class A specifications are identical with those detailed in paragraphs 4.2.2a to i inclusive for class B specifications.

**4.2.4.** Amendments to these specifications will be dealt with as proposals for new specifications according to the requirements detailed in paragraphs 4.1, 4.2.2, and 4.2.3 and their subparagraphs until incorporation into the relevant specification.

# 14

# Case Study 5: Control Procedure for Incoming Piece-Part Acceptance Specifications

Here again, as for Case Study 4, there had not been any critical problem. One or two minor hiccups had arisen, but generally, preparation of this procedure was one more step in the program for overall control of all the specification systems.

## A Procedure for Controlling the Preparation, Issue, and Revision of Incoming Piece-Part Acceptance Specifications

**1.** *Purpose.* This procedure is intended to ensure that piece-part acceptance specifications are prepared in a systematic manner and take into account the quality standards required of the company's end product. It is also intended to ensure that a control is exercised over the issue and revision of these specifications to prevent unauthorized use and revisions that could result in the acceptance of nonconforming material.

**2.** *Scope.* This procedure concerns primarily the engineering section of the QA department, who are responsible for the preparation and revision of piece–part acceptance specifications and for controlling their distribution. Specification Services (SS) is responsible for the control and issue of the specification masters. To a lesser extent, Production Engineering and Purchasing are interested parties.

**3.** *Policy.* There are to be piece-part acceptance specifications for all purchased parts and it is the responsibility of the QA engineering section to have them prepared.

**4.** *Related Documents*

BS 6001, ISO2859

MIL-STD-105D

**5.** *Procedure.* It must be noted that two quite different types of component are dealt with and the specification requirements for each are different. These component types are:

a. Mechanical

b. Electronic

They are dealt with separately in the following paragraphs.

**5.1.** *Mechanical Components*

**5.1.1.** For each individual component there is a detailed and dimensioned drawing from which it is made. The QA engineer responsible will discuss the various features shown on the drawing with the relevant product engineer. Jointly they will determine the relative importance of the features and the QA engineer will set acceptable quality levels (AQLs) and inspection levels.

**5.1.2.** Following on this cooperative action, the QA engineer will list the features in order of importance together with the AQLs and inspection levels and instructions about the type of inspection equipment that is to be used in each case. Depending on space requirements, the list is added to the master copy of the detail drawing, or it becomes a new sheet 2 of the drawing. The QA engineer must then advise the SS that the drawing is ready for issue.

**5.1.3.** The SS will issue the drawing to the approved circulation list and file the master copy.

**5.1.4.** The purchasing department is required to obtain from the SS up-to-date copies of the relevant piece-part drawing to send out with each enquiry to a prospective supplier or with each order.

**5.1.5.** When modifications are made to a piece-part drawing by product engineering, it is their responsibility to inform the relevant QA engineer of the detail of the change before the master copy of the drawing is amended. If the latter considers changes to the acceptance requirements to be necessary, he or she will discuss them with the product engineer. Both are required to approve all revisions. Taking account of the current purchase order position for the part, an implementation date will be set for introduction of the revision, and the purchasing department will be advised of this date. Otherwise, revisions are dealt with as in the case of new piece-part drawings.

**5.2.** *Electronic Components*

**5.2.1.** In the case of these components there is a procurement specification which sets out the requirements in each case. Many of these components are proprietary items.

**5.2.2.** The QA engineer responsible will consider each procurement specification and prepare a parallel acceptance specification bearing the same serial number. This specification will set out the tests that the engineer considers should be made to determine if the components meet the requirements of the procurement specification. As in the case of the mechanical components, the tests will be discussed and agreed to with the relevant product engineer and the AQLs and inspection levels will be set by the QA engineer.

**5.2.3.** Depending on the type and complexity of the component and the required tests, the acceptance specification will be sectionalized in the same manner as for "in-house" acceptance specifications prepared from customer specifications. The types of test equipment to be used will also be indicated.

**5.2.4.** When the specification master has been prepared, the QA engineer will send it to the SS for issue to the approved circulation list and for subsequent filing.

**5.2.5.** The purchasing department is required to obtain from the SS up-to-date copies of the relevant piece-part drawing to send out with each enquiry to a prospective supplier or with each order.

**5.2.6.** Revisions are dealt with as for mechanical components and described in paragraph 5.1.5.

# 15

# Case Study 6: Pointing to Specification Management

A number of events occurring over many years contributed to the development of the concept of *specification management*. The situations described in Chapters 7 to 12 were all problems that arose during my years in the field of quality management. The solutions were those actually developed and subsequently applied.

The solutions in Chapters 13 and 14 are in addition to those referred to above and have been included for two reasons: first, because they were developed and introduced by me, and second, because they completed a set of operational procedures for the control of all the types of specifications then in use by the company. In fact, although it was not realized at the time, they formed a basic system of specification management.

As time and effort expended on specification problems increased, it became apparent that much money was being wasted and that those losses were heavy and generally unsuspected — LOST COSTS in fact. The work done on Case Study 6 provided the first opportunity to carry out realistic costing. It was something of an eye-opener since, until then, no one had any idea of the significance and amount of the LOST COSTS attributible to specification problems.

This case study deals with a muddle in a specification system that resulted in a realization of the LOST COSTS. A major customer had introduced a new system of specifications for the electronic components that it bought in large quantities. There was a general specification for each class of component and a detail specification for each individual component in the class. My company was sent a parcel of between 40 and 50 specifications for four classes of components. Coupled with this was a request to quote for the supply of large quantities against each detail specification. There were four of the general specifications and the rest were detail specifications in the four classes. Although it had not then been fully developed, the procedure described in Chapter 8 was used as a basis for action in dealing with those specifications.

As four manufacturing groups were involved, the entire exercise was coordinated by the quality assurance department. There was a series of meetings with each product group and a final joint meeting at which the broad company attitude was agreed. The outcome was that the customer was asked to make 101 changes to the specifications in the parcel. Although not all of them were of the same significance, the customer was told that their specifications were not acceptable in the present form. They were also told that the company would not be agreeable to supplying components unless the requested changes were made. Of course, the customer was given full details of the changes being requested and the reasons for the changes.

The customer had ample time to consider the requested changes before a series of meetings lasting two days took place on the supplier's premises. The customer was represented by two specification/standards engineers who had been fully briefed on their company's attitude to the requested changes. In the course of the discussions all but one of the requested changes were agreed to by the two representatives. In that one case they said that they did not have full authority to agree to it and would have to discuss it further with their colleagues. However, they confirmed that each of the 100 changes that they had agreed to had been justified. They also agreed that it would have been better if more thought had been given to the formulation of the requirements in the specifications in the first place. Then the need for this series of major meetings, to say nothing of the internal meetings at both companies, would not have arisen. Some days later, a telephone call confirmed that the last change had been agreed to.

Although precise costs were not available, an estimate was made of the overall cost that the supplier company had incurred as a result of undertaking this exercise. Taking into account standard overhead charges and the fact that all personnel involved were indirect, it was estimated that at 1987 prices, the cost had been of the order of £20,000 ($35,000) and all in *indirect charges*. It was also "guesstimated" that the costs to the customer must have been of the same order. Not only did the customer have to revise all the specifications in the new system, but had to withdraw all those that had been issued to five prospective suppliers and reissue revised sets to them.

Some months later, during a visit, a very interesting and significant remark was made by one of the customer's engineers who had been involved in those discussions. He remarked that "it was rather funny but none of the other suppliers who had also received copies of those specifications had made any comment of any kind about them." He also wondered if any one at those other companies had actually read the specifications. If nothing else, this typifies the general disregard for specifications which seems to be so common. They lack integrity.

## How the Concept Developed

From all that happened during the preceding several years, the various problems that had occurred, and the solutions that had been developed, I came to the following conclusions. It was possible to identify a series of basic and major problems that were at the root of all the difficulties. It was obvious that those problems would have to be overcome if specifications were to be understood easily. These basic problems are now stated.

**1. The specification itself.** By its very nature the specification can cause problems, because there are so many things that can be wrong with it, and it arouses so many unhelpful reactions because of anticipated problems, whether they are there or not.

**2. Large organizations.** It is almost in the nature of things that large organizations cause problems. Because they are large, systems are necessary to ensure that everything is adequately controlled. Specifications also have to conform to the system. There is an excellent example in Chapter 18 of the harm that a system can do to a specification. When an organization is large, the controlling systems are also usually large and complex. In general, small organizations do not produce complex specifications. However, do not suppose that the small organization is free of the problems of the large one. More often than not it is the small organization that has to deal with the specifications of the large organization. It has the problem secondhand and that, if anything, tends to to make problem solving more difficult.

**3. The specification jungle.** It is well understood that people can get lost in a jungle, even when they have a guide. Specifications present a jungle to many people, either because they are inexperienced in their use or because the subject matter is not familiar to them. Or because the language in which the specification is written is too involved for them to understand. This jungle is a prime outcome of problems 1 and 2.

**4. Specification interrelationships.** Many specifications are intended to be used on their own without reference to another specification. However, many more do require reference to one or more additional specifications. This is a frequent cause of confusion, as shown by some of the examples in Chapter 5. There is also insufficient realization on the part of many specification writers of the problems that can arise when reference to other specifications is not crystal clear.

**5. Effects of specifications on quality standards.** Generally, people do not realize that without specifications there cannot be any product with any degree of quality. In fact, without specifications of one kind or another, there cannot even be a product. Unfortunately, very few people realize that the quality of a product depends on the quality of the specification itself. That factor, the quality of specifications, is what this book is all about.

**6. Variations in specification-writing competence.** Remember the statement that appeared at the beginning of Chapter 7: "I know that you think that you understand what I said. But what you do not understand is that what I said is not what I meant!" It must be said that there are many specifications of high quality. Unfortunately, however, there are also many specifications whose quality is more than suspect. There are many reasons for this variability, and some of them relate to the size of the organization responsible for the specifications. In too many cases, the writers of specifications are too remote from the people who will be using them.

In many cases it seems as if the writers have little or no patience with the needs of those who need to understand what they have written. There are also the pressures that are exerted on specification writers from within their own organizations—quite often those of the marketplace (e.g., the urge to have a product on sale before a competitor does). This adds to the very real difficulties that many engineers have in expressing their requirements clearly and unambiguously.

**7. LOST COSTS.** This is, perhaps, the biggest problem of all. The LOST COSTS have been mentioned many times. They warrant many references, as, in total, they amount to a very large sum of money for industry as a whole. In a British government consultative document published toward the end of 1978 (*A National Strategy for Quality;* Her Majesty's Stationery Office) it was stated that the total cost of quality to British industry was in the order of $20 billion annually. No one knows for sure, but it is likely that the proportion of this huge annual loss which is caused by specification problems is quite significant. Even if only 1% of it could be attributed to this cause, it would still be as much as $200 million annually (see the discussion on the cost of specification reading time in Chapter 16.)

In Case Study 6, $35,000 was spent to prevent later losses, which would have been much higher when all the interrelated and consequential actions were taken into account. Yet, had the necessary effort been expended in the first instance, the cost would have been much less than $35,000. Many of the losses (LOST COSTS) said to be the fault of machines, operators, and materials should really be laid at the door of the specification writer. When a requirement is not properly understood or is confusing or needs to be interpreted, it is usually much easier to blame the operator rather then the specification. And that is what usually happens.

**8. Frustration.** Frustration is an all-too-common human ill, and not the least of its manifestations arises as a result of the psychological effect often resulting from the use of poor-quality specifications. Because of difficulty in understanding, in interpretation, or because of duplicated or confusing statements, readers will often say a few rude words and discard it in total frustration. It will be less frequently consulted in the future, if at all, and that will cause more problems.

**9. Deteriorating relationships.** When one is frustrated, there is a tendency to be "short" with colleagues, and relationships that ought to be friendly will suffer. This is of considerable importance in industrial and commercial situations. Any relationship into which strain enters will not work as well as one in which there is no strain. All kinds of things are likely to be said and done which in the long term will do nothing but harm to the product because of the mistakes that will arise.

**10. Specification integrity.** This problem is apparent as a disregard for what a specification says. It is not as rare a phenomenon as it should be and is a natural outcome of some combination of the previous nine problems listed. The specification ought to be the bible for its users. Everything possible should be done to restore the firm belief among users of specifications that the specifications are right.

This chapter has been concerned with the problems I have encountered in my search for an overall solution. The next chapter deals with the result of that search, the concept of specification management.

# 16

# The Concept of Specification Management

Few, if any, company managements would dream of operating without a fire insurance policy. The risk of loss would be far too high. On the other hand, few, if any, managements operate any policy to reduce the (LOST) costs caused by the seldom-suspected problems arising from poor-quality specifications.

The concept of specification management provides just such an insurance policy. It can be established as a valuable tool to control and reduce the undesirable hidden expenditures resulting from the use of poor-quality specifications—expenditures seldom controlled at all and seldom appreciated. The recovery of LOST COSTS attributable to specifications, as well as notable improvements to the technological image that a company presents to the outside world, can result directly from the adoption of this concept. Perhaps of even greater importance, it can effect a substantial reduction in the number of quality and reliability problems with which industry is bedevilled.

There are three British Standard publications which are intended to provide guidance to writers of standards and specifications:

BS 0        A Standard for Standards
PD 2879     Drafting Guide for Electrical Standards
PD 6112     Guide to the Preparation of Standards

Unfortunately, these BS guides are of little help in the writing of specifications or standards which are to be clear and unambiguous, unlike the examples discussed in earlier chapters. Although PD 2879 does devote one of its 22 pages to a small measure of guidance, BS 0, which had recently been revised, is a little better—but not much. None of them deal with the kind of faults that occur so frequently and contribute so often to the misunderstanding of specifications. A particularly important point in all these BS publications is the omission of examples of right and wrong. This is singularly unfortunate since there is a considerable body of experience to confirm that the use of practical examples does provide a very simple and effective way of making a point. It may also be of some significance that I have never seen any of these guides in use.

It is appreciated that some organizations do have an editing function that operates with a varying degree of success. However, it is much better to try to create an environment in which the basic material is prepared in a style and format that requires the minimum of further attention. It is also much cheaper. It should be remembered that in such circumstances, no editor can be expected to be expert in all of the disciplines and techniques covered by the specifications that he or she has to handle.

Many departments in a large organization are concerned with specifications and procedures. Therefore, the system of management suggested should operate across the organization on a general service basis in much the same way that accountants do. For the same reason, the manager or controller of the new service should, of necessity, be a senior manager in the organization.

The basic responsibilities of the specification manager, which for obvious reasons will not include responsibility for the actual technical content of the specifications, would be as follows:

1. To be responsible for overall editorial control to ensure that specifications are written in a uniform, simple, easy-to-understand style
2. To ensure that presentation of the content is logical, clear, and unambiguous
3. To ensure that uniform standard formats are used which recognize the different purposes to which specifications are put

4.  To ensure that cases in which it is clear that there will be consequential actions arising from the technical requirements they will be fully appreciated by the originator of the specification
5.  To ensure that there are logical and coherent numbering and titling systems and suitable procedures for the issue and revision of the specifications.
6.  To ensure that interested parties within the organization, but outside the immediate circle of the orginator, are given the opportunity to express their views about the proposed content of specifications while they are still "proposed" and readily changeable with minimum cost
7.  When it would seem advantageous to do so, to ensure that opinions are sought from parties outside the organization, especially should the views of suppliers be sought when specifications relate to possible purchases

Many of the problems that will be met by a specification manager have already been discussed in some detail, so it is clear that his or her task will not be an easy one. There are two additional problems which are likely to cause much difficulty. Many of the disciplines in manufacturing organizations are noted for their jealous adherence to the usages and practices established over many years. This fact, together with the usual resistance to change—the inertia—normal in any organization, especially the larger ones, will necessitate a specification manager with a high degree of tact and diplomacy.

Experience has shown that specification writers do not always take kindly to suggestions that they should change this, or the ordering of that. They are even less likely to look kindly on suggestions that some of the fruits of their toil could cause financial loss to their employers. On the other hand, there are writers who will appreciate and welcome the assistance that a specification manager can provide. They recognize their own limitations or, often more important, the artificial limitations imposed by administrative or outside pressures. The case study that is dealt with in depth in Chapter 18 is a case in point. Marketing pressures are yet another example of these limiting factors.

In a conference that I attended in London many years ago, a paper was presented on the subject of miniature relays. During the discussion that followed, one member of the audience put a question that seemed to point to one of the many reasons for specification problems. "Would the speaker please explain why it is that I, as a user of miniature relays, have to do so much of the development work which ought to have been done by the manufacturer? Nor does this criticism apply only to your product. I have found the same problem with most manufacturers!"

The answer from the presenter of the paper was simple and straightforward: "No doubt it is my fault, but I allow myself to be 'persuaded' by my marketing colleagues that time is of the essence. That if our product does not come on the market 'tomorrow' we will have lost out to our competitors and all our work will have been wasted. Of course, this is not really the case, but!"

The discussion continued for a short time on the need for product designers to resist marketing pressures so that a product should not go on to the market until the designer is satisfied that it is right. This does not necessarily mean that the product is perfect initially, but it should mean that it is at least workable. Further improvements will come with time. This is an important point that a specification manager must keep in the forefront of his or her thinking. It is not the specification manager's job to ensure a perfect product. But it will be his or her task to see that the specifications for the product are clear and understandable. If the designer wishes to produce specifications for square pistons, that is up to the designer. But the specification manager must ensure that those specifications for square pistons *are* clear and understandable, so that, in fact, square pistons can be made from them without difficulty. Of course, the specification manager will also have to ensure that the designer is fully aware of the consequences of designing square pistons.

Another entrenched difficulty that the specification manager will have to face in some large organizations is that of procedure systems—procedures that have been in existence for a long time and have become hallowed and unchangeable. The procedure may prescribe that the subject matter in a specification must be in a certain order—that certain kinds of information in certain types of specification must always be included even if only to say "not required." The specification in Chapter 18 is a case in point. It was required that a large number of specified items be referred to in a particular order whether required to be or not. The first version of this specification shows the considerable waste of paper caused by all this redundant material, and the wasted time of the reader and user. A drastic reduction in both paper and time followed the production of the revised version (also given in Chapter 18). Following the work done on this specification, it is understood that that system has been changed to permit the preparation of considerably simplified specifications.

That example also shows how a specification manager could make clearly identified cost savings. Savings resulting from a substantial reduction in reading time. How many people would realize that the cost to a large organization for someone to read the original version of that specification could be as much as $10. By reducing the amount of reading matter, in this instance by about 75%, the cost saving per reading could be as much as $7. If one also takes into account the time needed to understand a specification, it is likely that the time required would considerably exceed the simple reading time in most cases—all adding to those unsuspected and considerable LOST COSTS. Reverting to the example, if one considers the number of times that this specification alone will be read *and* understood during its lifetime, it is not difficult to see how the sum of $200,000,000 referred to in problem 7 in Chapter 15 arose.

It is important for specification managers to remember those LOST COSTS, since the organization's losses in time and money will be reflected in the costs of suppliers. One might say that this is the suppliers' problem. But do not forget that whatever cost a supplier may incur, it has to be paid by someone, and that someone is always a customer, although customers seldom, if ever, realize that they are paying for this LOST COST. In addition to the specification manager, it is important for others to remember this point: *It is always the customer who pays in the long run.*

*Every unnecessary minute that it takes to read and understand a specification is likely to cost the reader's employer upward of $0.35.* Perhaps that reason, rather than all the strictly ethical ones, is the one that will carry most weight in an organization when considering whether or not to introduce a system of specification management. It will save time for *all* the readers and users of specifications—and *time is money.*

Although to a large extent the various discussions that have appeared in this book so far have, at least by implication, assumed that the readers/users of specifications are all in supplier companies, this is by no means the case. The circumstances apply with equal force to the internal use of specifications in large organization. Here it is perhaps easier to demonstrate the extent and magnitude of LOST COSTS. The example described in Case Study 6 in Chapter 15 emphasizes the point.

In the example described in example 3.1.b in Chapter 6, considerable LOST COSTS were incurred. It was estimated that, at 1987 prices, they were of the order of $1000. A little more thought from the originators of the requirements in the first instance would have saved most of that $1000.

The most obvious result of the introduction of specification managment is not, however, the saving in the LOST COSTS. It is the considerable improvement in the quality of the specifications that would result, and the reduction in the number of problems arising from them. But the most important result is the savings in LOST COSTS.

# 17

# Introducing Specification Management

As the way in which the task of introducing a system of specification management will depend on a variety of factors, including the size of the company or organization, only some points of general guidance can be given. With them, and the information in the earlier chapters, it should be relatively easy to plan suitable action for any size of company or organization and complexity of specification system. However, it is best that in any case the introduction should be done on a gradual basis.

Once the decision has been taken, the choice of a title for the "controller" will depend to some extent on the size of the organization. Also, in a sense, the person's technical qualifications are not critical. He or she must be literate and numerate and have plenty of common sense. Experience in writing, and possibly in technical editing, would be very helpful, although by no means essential. A general knowledge of the technology(ies) dealt with in the organization is also desirable (perhaps some knowledge of "square pistons").

The requirements for a smaller company will be considered first. It is quite likely that at least in the first instance, the job can be carried out on a part-time basis. An initial task for the controller (this is a suitable title for this person in a small organization) will be to familiarize himself or herself with the various specification systems currently in use. There are likely to be at least four for manufacturing, final acceptance, incoming material and customer. They could be dealt with in that order unless there are local reasons for an alternative order. That is not a matter of great import.

Whichever specification group is selected for a start, a careful study must be made of it. There should be discussions with those who use the specifications as well as with those who write them. From the discussions and the study it will be possible to develop a procedure that will cover all aspects of preparation, issue, use, and revision of the specifications. It is possible, or even likely, that changes will be considered to existing numbering and titling systems and, of course, formats. Guidelines will have to be prepared for the writers. They should include a variety of information, some of which has already been mentioned, but not forgetting guidance about ordering of the contents of specifications. It is important that any changes that are made to existing procedures should have been discussed with all concerned and, if at all possible, accepted by them before formal implementation. Once the first group has been brought under control, the rest can be dealt with in turn. This, incidentally, was the method used in the case studies of the earlier chapters. They will, of course, provide sound guidance.

In the case of large organizations the situation will be quite different, as among other things, it is quite likely that there will be a number of specification writing groups which are almost certainly independent of one another. In this case it would be appropriate for the title of the controlling person to be "specification manager." The manager's first task should be to make a detailed survey of all the specification systems in use and the controls, if any, that govern their preparation. The outcome of such a survey is sure to bring some surprises. It will almost certainly be discovered that there are more specification systems, and more controls over them, than was thought to be the case. Systems may be found which are nearly identical but operated on quite different arrangements! Some of the controls may be of a purely administrative nature and take more account of the convenience of the administrators than of those who have to use the specifications. Other controls are likely to be out of date. It will almost certainly be found that there are as many ways of presenting specifications, formats, and the like, as there are originating departments— perhaps even more. It will be surprising if any of the formats take into account the needs of the users of the specifications.

In any event, a great deal of information will be gathered and the manager will have to spend some time considering it. It will be neither a quick nor an easy task. The manager will have many discussions with the various groups involved and in these discussions will need all his or her reserves of diplomacy. In due course, however, the "murk" will begin to clear and the manager will be able to begin planning for the formal introduction of the total system for specification management. There will probably be a number of parallel streams identified with the various departments or groups producing specifications. It is desirable that for the first group, at least, the manager should do as much of the work personally as possible. This will help to set the general targets for which he or she is aiming and provide further guidance for those who will follow.

As soon as the first group is working on the right lines, the manager should begin recruiting assistance to take over the parallel streams one at a time. This should not be done too rapidly or there is a risk of losing control. The manager's recruits will have to be indoctrinated and gently slipped into their allocated stream. Supervision will have to be closely maintained for some time. The manager must remember that he or she is unlikely to be able to recruit experienced persons. It is a new kind of task; the recruits will have to be trained on the job.

On the question of time scales. Much will depend on a variety of factors, not the least of which is the way in which the manager sets about the task: whether he or she develops guideline documents first or carries on that work in parallel with other tasks. Other factors are the size and nature of the organization; the general nature of the plan the manager prepares; and the willingness of the manager's various colleagues to work with him or her. It will be a slow process and this fact must be fully recognized. There is also the NIH (not invented here!) syndrome, and at times the manager might well feel like throwing in the towel. But the prizes to be gained are considerable and worth any "angst" that might lie in the way. In the case of the work done on the case studies, which was all part-time in the first instance, the time from the beginning of the first exercise to the last was between 18 months and two years.

It is likely that in practically any type of company a considerable part of the LOST COSTS can be saved with a specification management system in full operation. This saving could be as much as 1%

of the annual sales income of the company, thus freeing much time and effort for more profitable activities. It is realized that only in manufacturing companies can the savings be expressed in this way, but in other organizations it should not be difficult to find similar markers.

# 18

# Case Study 7: A Recent Practical Exercise

This chapter contains the complete details of a recent practical exercise making use of the principles in this book. A 16-page specification that was difficult to read and understand was taken as an example. The difficulties it caused were due to a number of reasons of which the main ones were the following:

1. There was much redundant information.
2. There was much duplication of information.
3. The ordering of the contents was very disjointed.
4. There was no index.
5. The format was dictated by administrative procedure.
6. Problems 1, 2, and 3 were due entirely to problem 5.

As a result of the exercise, a revised version of the specification was produced with only four pages, plus two additional pages for appendices to contain the unnecessary information demanded by the administrative procedure. The revised version contains all the essential information from the original specification, and it is much easier to read and understand. The bare reading time for the original was approximately 18 minutes, without allowing for the additional time necessary for full comprehension. The reading time for the revised version was about 5 minutes, with a much reduced comprehension time. On the basis of the figures given in

Chapter 16, this represents a saving in reading time costs of about $7 for every reading. The savings in comprehension time are likely to be much greater.

In the following pages will be found, in this order:

Page 144: The text of the original specification, annotated with initial comments on the original specification
Page 159: A tentative for the original specification
Page 166: The revised version of the specification

It is of interest to note that after this exercise had been completed, as part of a training course on specification writing and management, the organization concerned decided to make changes in its administrative requirements for the preparation of specifications. These changes resulted in much shorter and simpler specifications and substantial cost savings fewer LOST COSTS.

## TEXT OF A FUNCTIONAL SPECIFICATION FOR THE PROVISION OF TOXIC MATERIAL CONTAINER MAINTENANCE WORKSHOPS*

*4.1.1.*     **1.** *Brief description of the building.* The building is mainly a single-story building, but incorporates an extract plant room and a tank room at the first-floor level. It has concrete floors, brick walls, and a concrete roof. Ventilation is provided, as are lights and background steam heating.

This specification relates only to modifications required in the two rooms, 8 and 9.

*4.1.2.*     **2.** *Function of the building.* Remainder, no change. Rooms 8 and 9 are to be used for the maintenance of the containers utilized for the transport of toxic materials.

*In this paragraph the information seems to be in a rather illogical order. It would be more logical to indicate the functions of rooms 8 and 9 before indicating that there is no change in the remainder. In any*

---

*Incorporating Modifications to Structure and to Services of Existing Buildings, Specification Reference Z 100 *event, the information is partially repeated in para-graph 2.5.*

App. 2    **2.1.** *Explosives.* Nil.

4.8.     **2.2.** *Toxic material.* Not normally expected, but a small amount of toxic material may be expected during servicing of the containers.

4.1.2,   **2.3** *Layout of rooms 8 and 9.* Layout of the rooms is
4.1.3    to provide:

Reception and services area (part of room 8)
Servicing area (room 9)
Personnel and monitoring area (part of room 8)

4.2.1    **2.4.** *Access to building.* Access will be required from the roadway existing at the eastern end of the building, the access to cater to:

Personnel
Vehicles, delivering and collecting material containers

*In paragraph 2.4 there is no information about load-bearing requirements for the concrete hardstanding. With vehicular use, this should have been supplied.*

4.1.2    **2.5.** *Rooms 8 and 9, function.* The rooms are required to provide handling and storage space for toxic material containers. Services are to be provided to maintain a suitable environment during normal conditions.

4.9     **2.5.1** *Fire certification.* CFO to arrange application for modification to existing certificate or new certificate, as appropriate.

App. 1   **2.6.** *Special safety precautions.* Adequate ventilation of the two rooms will be necessary as specified in section 22.

*This information is repeated, generally, in paragraph 22. Then, in 22.2 and 25.11 it is said that there is no special requirement.*

4.8     **2.7.** *Toxic solid wastes.* The approved toxic material containers to be serviced will be certified "clean" before entry into the building, but minor amounts of toxicity may be exposed during dismantling and cleaning of the containers.

*This information seems to be at least slightly con-tradictory, as one would expect that any item cer-*

*tified "clean" would be "clean." And paragraph 22.2
suggests that precautions are not required.*

*4.12*       **3.** *Numbers of occupants.*

    **3.1.** Male:   Industrial    As required (4
                                     maximum)
                     Nonindustrial  As required (2
                                     maximum)

    **3.2.** Female:  Industrial     Nil
                     Nonindustrial  Nil

    **3.3.** Ancillary staff: Male. The normal ancillary sup-
port staff, industrial and nonindustrial, will be enter-
ing the building as for other buildings in the area.

    **3.4.** Ancillary staff: female. As paragraph 3.3.

*App. 2*      **4.** *Accommodation for ancillary services.* None required.

*3.1*         **5.** *Associated drawings.*

    Modified building plan
    Modified electrical layout
    Modified mechanical layout

*App. 1,*     **6.** *Roofs*
 *App. 2*    **6.1.** *Construction.* No change.

    **6.2.** *Access walkways.* Nil.

    **6.3.** *Load bearing.* No change.

    *In paragraph 27 there is a requirement for the in-
stallation of two 1-metric-ton hoists. However, there
is no information in any part of the specification
about their installation. One imagines that there
would be some structural alterations required to ac-
commodate them. These should have been given
here.*

*App. 1*      **6.4.** *Thermal insulation.* No change.

*4.5.1,*      **7.** *Floors*
 *App. 1,*   **7.1.** *Construction.* No change.
 *App. 2*    **7.2.** *Finish.* Rescreeded to renew surface and
covered with linoleum.

    **7.3.** *Dustproofing.* None.
    *This is quite redundant information. Paragraph 7.2
should have been enough.*

    **7.4.** *Loading.* No change.

**7.5.** *Mat wells.* Nil.

**7.6.** *Timber.* None required.

*App. 2*     **8.** *Ducts and drainage channels.* None required.

*This information would seem to be in conflict with paragraphs 33.2 and 30.7.1 if their titles are to be taken as written.*

*4.5.3,*     **9.** *Walls*

*App. 1*     **9.1** *Construction.* Existing lintels and brickwork above door opening between rooms 8 and 9 to be removed, and after completion of all new work all finishes are to be made good to existing standards.

**9.2.** *Finish.* No change.

**9.3.** *Acoustic.* No problems anticipated.

**9.4.** *Paint.* To be repainted using chlorinated rubber paint.

**9.5.** *Air bricks.* No change.

*4.5.2*     **10.** *Ceilings*

**10.1.** *Finish.* To be repainted using chlorinated rubber paint.

*Here "Finish" calls for painting. In paragraph 9.2 it is "No change" and in paragraph 9.4 it is "Paint"; the requirement is the same as in paragraph 10.1, "Finish."*

**10.2.** *Irregularities.* To be kept to a minimum to faciliate cleaning of surfaces.

*4.4,*     **11.** *Doors*

*App. 2*     **11.1.** *Construction.* See separate drawings of doors.

*When these separate drawings were produced they were found to be location drawings only, not construction.*

**11.1.1.** *Existing door No. 65.* Entrance door to room 8 to remain unchanged.

**11.1.2.** *New door (No. 70).* The window in the east wall of room 8 to be removed together with frame and all brickwork to floor level and the lintel and brickwork over the window. Replace with 75-mm-thick purpose-built timber-faced doors 2828 mm high × 1970 mm wide overall, including frame. Doors to be trimmed around monorail.

*The structural changes here should have gone into paragraph 9.1, which deals with other work on a wall.*

**11.1.3.** *Door alteration, No. 5.* Door between room 9 and building internal corridor to have blanking boards removed and 50-mm-thick half-hour-resistant timber-faced doors fitted into existing opening. No glazing to be fitted to any doors.

*It would be more logical for the information about "no glazing" to appear in paragraph 11.1. As it is in any case not required, why bother?*

**11.2.** *Self-closure*

**11.2.1.** *Existing No. 65.* No change.

**11.2.2.** *New door No. 70.* Not to be self-closing.

**11.2.3.** *Door alteration, No. 5.* To be self-closing (Briton 600P floor closer unit).

**11.3.** *Electric alarms and locks*

**11.3.1.** *Electric alarms.* Nil.

**11.3.2.** *Locks*

**11.3.2.1.** *Existing No. 65.* No change.

**11.3.2.2.** *Door No. 70.* Mortice lock to be fitted, with key mastered XLA.

**11.3.2.3.** *Door No. 5.* Mortice lock to be fitted—lockable from passage only.

**11.4.** *Fire resistance.* All doors to have half-hour fire resistance.

*This paragraph and the next seem to be an example of instant duplication.*

**11.5** *Non-fire-resistant doors.* None to be fitted.

**11.6.** *Plant room doors.* Nil.

*Oddly enough, there is a plant room on the first floor which is not affected by the alterations.*

**11.7.** *Internal locks*

**11.7.1.** *Door alteration, No. 5.* Lockable from passage only, as stated in paragraph 11.3.2.3, and to be provided with two keys, each key to have 1½-inch (32 mm)-diameter steel ring with 1½-inch (32 mm)-diameter brass tablet stamped with the appropriate door number.

*Except for the additional information about the*

*keys, this item has already been dealt with in paragraph 11.3.2.3, in which the key details could have been given.*

**11.8.** *Laboratory peeps.*   Not applicable.

**11.9.** *Emergency exits.*   Not considered necessary. *This is at least the fourth word phrase that gives the same meaning.*

**11.10.** *Cabin hooks*

    **11.10.1.** *Door No. 70.*   To have two off-garage stays.

*This information would seem to be more appropriate in paragraph 11.1.2.*

*4.3,*     **12.** *Windows*

*App. 2*     **12.1.** *Position, size, type.*   The window in the middle of the east wall to be removed to provide the main new door. All other windows to remain unaltered.

*Although in this case there is no reference to the room in which it is, this has already been dealt with in paragraph 11.1.2.*

**12.2.** *Fixed/horizontal/vertical.*   Not applicable.

**12.3.** *Single or double glazing.*   Not applicable. *This has, in any case, already been dealt with in paragraph 11.1.3.*

**12.4.** *Special requirements.*   Not applicable.

**12.5.** *Electrical substation.*   Not applicable.

*4.4*     **13.** *Height of doors.*   No change.

*Quite why this item is here on its own instead of in section 11, which deals with doors, is a puzzle.*

*App. 1*     **14.** *Anti-vegetation barrier.*   No change.

*4.2*     **15.** *Road approaches*

    **15.1.** *To main access door No. 70.*   Provide concrete hardstanding to new door from existing road on east of building.

*This could have been dealt with in paragraph 2.4, which deals with access to the building.*

    **15.2.** *Car parking and cycle racks.*   None required.

*4.6.1*     **16.** *Special fitments*

    **16.1.** *Change area equipment.*   Mobile shoe barriers, three sets of five coat hooks, and two overshoe bins to be provided. Two wash basins, water heater, and hand drier also required in room 8.

*App. 1,*     **17.** *Toxicity safety*

*App. 2*      **17.1.** *Boundary walls.* No change.

**17.2.** *Entrance.* No special facilities required.

**17.3.** *Storage facilities.* No special storage facilities required.

**17.4.** *Detectors and alarm systems.* Not required.

**17.5.** *Procedures for detectors and alarm systems.* Not applicable.

*4.9*         **18.** *Firefighting appliances*

**18.1.** *Portable appliances.* Firefighting appliances will be specified by, and supplied by, the chief fire officer.

**18.2.** *Permanent equipment.* $CO_2$ or other permanent equipment is not required.

**18.3.** *Fire control box.* No change.

*4.9,*        **19.** *Fire warning systems and evacuation route*

*App. 2*      **19.1.** *Automatic fire detection in room spaces.* Not required.

**19.2.** *Automatic fire detection in apparatus.* Not required.

**19.3.** *Fire evacuation alarms.* Not required.

**19.4.** *Fire evacuation routes.* Not applicable.

**19.5.** *Emergency exits.* Not required.

**19.6.** *Layout of plant.* Not applicable.

*4.10*        **19.7.** *Building classification symbols.* The building classification symbol and its location shall be in accordance with local orders and shall be determined in consultation with the chief fire officer.

*App. 2*      **20.** *Handrails or staircases, access ladders, maintenance walkways.* None required.

*This fact is also confirmed in paragraph 27.3.*

*App. 1*      **21.** *Heating*

**21.1.** *Type of heating.* Steam.

**21.1.1.** *Existing heating.* Some alterations to the sitting of the existing steam pipes may be necessary to reroute around new lifting beams and doors. Generally, the system will be as existing.

*Presumably the lifting beams are the monorail*

*hoists, which from the description, do not seem to be the same things.*

*App. 2*   **21.2** *Radiators.* Not applicable.

*App 1*   **22.** *Ventilation.* Ventilation is required to cater to any minor exposure to toxic material that may occur during the servicing of the containers.

*This is in some conflict with the next paragraph, as the rooms were not, apparently, previously affected by toxic materials.*

**22.1.** *Type of ventilation.* Existing filtered extract system to be used, which would give up to 20 room changes per hour.

*App. 2*   **22.2** *Special safety requirements.* None required.

*See comment on paragraph 2.6.*

*App. 1*   **22.3.** *Damper requirements.* No change to existing.

**22.4.** *Plenum ventilation.* Existing system to be used.

**22.5.** *Extract ventilation.* Existing system to be used.

*Already covered by paragraph 22.1.*

*App. 2*   **23.** *Air sampling.* None required.

*4.6.2.7*   **24.** *Lighting*

**24.1.** *Type and intensity* Mains-fed fluorescent lamps complying with British Standards are to be used throughout. Intensity of illumination in room 9 at floor level shall be not less than 300 lux and not less than 150 lux in room 8. Light switches to be grouped on the north (right) side of the opening between labs 8 and 9.

*App. 2*   **24.2.** *Special requirements.* None required.

*See the following paragraph.*

**24.3.** *Emergency lighting.* None required.

**24.4.** *Cranes.* Not applicable.

**24.5.** *Bench lighting.* Not applicable.

*App. 1*   **25.** *Internal electrical services*

**25.1.** *Internal power requirements*

*4.7*   **25.1.1.** *Single-phase ac.* The existing single-phase 240-V 50-Hz service is sufficient to supply as follows:

All lighting
2M panels or equivalent in room 8
4 × 5 ampere socket outlets in room 8
2 × 5 ampere socket outlets in room 9
1 × 15 ampere switched outlet in room 8 for hand drier
Supply to water heater
Clock supply
PA unit

*The socket outlet requirement is repeated in paragraph 25.6.1. It is only audio lines that are required for the PA system.*

*App. 2*  **25.1.2.** *Three- phase ac.* None required.

**25.1.3.** *Dc voltage supplies.* None required.

**25.1.4.** *Essential supplies.* None required.

**25.1.5.** *Stabilized voltages.* None required.

*4.7*  **25.2.** *Earthing.* All earthing to conform to standard. The earth connection is that of the building incoming cable.

*App. 2*  **25.3.** *Standby services.* None required.

*4.7*  **25.4.** *Location of sockets*

*They do not actually appear until paragraph 25.6.1 under "Miscellaneous; Portable tools," and with no more information than already given in paragraph 25.1.1.*

**25.4.1.** *Hand driers.* One hand drier is to be provided in room 8.

*This hardly gives the location of the socket.*

**25.4.2.** *M panels.* M panels or their equivalent are required, one on the north wall and one on the south wall of room 9.

*Not really "sockets" as intended by the use of the word.*

*App. 2*  **25.4.3.** *Essential power supplies for alarm systems.* Not required.

*Already covered in paragraph 25.1.4.*

**25.5.** *Extra low voltage.* Not required.

*4.7*  **25.6.** *Miscellaneous*

**25. 6.1.** *Portable tools.* Four sockets required in room 8. Two sockets required in room 9.

*4.6.2*       **25.6.2.** *Water heaters.* One water heater to be provided in room 8.

*In paragraph 29.9 both wash basins are paid to be located in room 9.*

*4.7.2.*      **25.7.** *Earth leakage trip system.* An earth leakage trip to be provided in the supply to the DFB, supplying all the outlets called for in paragraph 16.1.

*App.1,*      **25.8.** *Dusttight and flameproof installation.* Not re-
*App. 2*   quired.

**25.9.** *Overhead busbar system.* Not required.

**25.10.** *Conductive flooring and screening.* Not required.

**25.11.** *Special safety requirements.* None required.
                    *See comment on paragraph 2.6.*

**25.11.1.** *Individual plant isolation.* None required.

**25.11.2.** *Distribution board location.* No change.

**25.11.3.** *Services needing essential supplies.* None required.

*Paragraphs 25.1.4 and 25.4.3 already say that essential supplies are not required.*

*4.7*       **25.12.** *Final circuits.* All final single-phase circuits are to be served from TP and N distribution boards. DFB 9 × E power supplies, DFB 1B lighting.

*It is not the least clear what the relationship is between the M panels of paragraph 25.1.1 and the ones referred to here.*

*App. 2*   **26.** *Controls and alarms*

**26.1.** *Alarms/indicators, types.* None required.
*Paragraph 17.4 has already said that none are required.*

**26.2.** *Control and alarm panels*

**26.2.1.** *Separation.* None required.
               *Unnecessary statement after paragraph 26.1.*

**26.2.2.** *Fire alarms.* None required.
               *Unnecessary statement after paragraph 26.1.*

*4.6.2.5*   **26.2.3.** *Public address.* One PA speaker to be positioned on the north wall in the northwest corner of room 8.

*App. 2*   **26.2.4.** *Battery charger.* None required.

**26.2.5.** *Glove box alarms.* Not applicable.

**26.2.6.** *Control building.* None required.

**26.3.** *Intercom system.* None required.

4.6.2.4    **26.4.** *Electric clocks.* One clock is required, located in the center of the east wall at high level.

*By deduction from paragraphs 3.3 and 12.1.2, this suggests that the clock is to be in room 9.*

4.6.2.6    **26.5.** *Telephones and public address system*

**26.5.1.** *Telephones.* One telephone is required in room 8 near door No. 70.

**26.5.2.** *Public address units.* One PA speaker located as in paragraph 26.2.3.

*Why bother to say it again?*

*App. 2*    **26.6.** *Mess room.* Not applicable.

4.6.2.8    **27.** *Cranes, lifts, hoists, and monorails.* Two monorails required, one in room 8 and one in room 9.

**27.1.** *Lifting capacity.* Each monorail to have a lifting capacity of 1 tonne.

*The requirement in paragraph 39.2 would have been better here.*

**27.2.** *Electric/hand operation.* Hand hoist operation.

*App. 2*    **27.3.** *Access/walkways.* Not applicable.

*Paragraph 20 has already said this.*

**27.4.** *Miscellaneous*

**27.4.1.** *Roof clearance.* Not applicable.

*App. 2*    **27.4.2.** *Emergency stop.* Not applicable.

4.6.2.8    **27.5.** *Special safety requirements.* 12-mm plate to be welded at each end of monorail in room 8 as stops.

*App. 2*    **28.** *Compressed air.* None required.

*App. 2*    **29.** *Effluent and drainage*

*This main paragraph and 30.7 deal with the same subject, with at least three "kinds" of water in each.*

**29.1.** *Sewage (foul drain).* Not required.

4.7.7    **29.2.** *Toxic material drain.* The water from the wash basins should flow by gravity to the sump tank via the toxic drain system. Flow not expected to exceed 50 liters per day.

*In this paragraph and 30.2 there is a mixup about the quantity of water per day. Here it is given as 50 liters and in the later paragraph it is given as 20 gallons. These quantities are quite different.*

App. 2     **29.3.** *Trade waste.* Not required.

App. 1     **29.4.** *Storm water.* No change.

App. 2     **29.5.** *Photographic waste.* Not required.

4.8     **29.6.** *Toxicity level of drains.* No change from existing level of toxicity of drains.

4.7.7     **29.7.** *Pipes.* To be stainless steel or approved plastic.

**29.8.** *Pipe details*

*The information in paragraph 29.7 should be here.*

**29.8.1.** *Accessibility.* No change to existing pipework. Any new pipes to be accessible throughout their entire length.

**29.8.2.** *Leakage.* No change to existing pipework. Any new pipes to be arranged so that leakage is collected.

*Nowhere is one told how the leakage is to be collected.*

**29.8.3.** *Self-draining.* No change to existing pipework. Any new pipes to be self-draining and free from possibility of blockage and/or air locking.

*Three paragraphs in succession have said there is "no change to existing pipework."*

4.6.2.1,     **29.9** *Wash basins.* Two wash basins required in
4.7.6     room 9 discharging into the drainage channel via stainless steel or approved plastic strainers and traps. Taps to be elbow operated.

**30.** *External services*

App. 1     **30.1.** *Types and details*

*Steam:* Existing service is adequate.

*Condense:* Existing service is adequate.

*Water:* Existing service is adequate.

App. 2     *Gases:* None required.

App. 1     *Electricity:* Present supply adequate.

*Drains:* Existing service adequate.

4.6.2.6     *Telephone:* One required.

App. 2     *Compressed air:* None required.

*4.2.1*    *Roads and paths:* Modify access at eastern end of building to improve vehicular access to entrance from existing road.

*Several of these items have been dealt with in previous paragraphs.*

**30.2.** *Loads*

*App. 1*    *Steam:* No change.

*Condense:* No change.

*4.7.5*    *Water:* 50 liters per day.

*App. 2*    *Gas:* None required.

*4.7.1*    *Electricity:* To supply the requirements of paragraph 25.1.1.

*4.7.7*    *Drains:* Connection to sump tank to take 20 gallons per day by gravity from wash basins.

*4.6.2.5,*    *Telephones:* 2 pairs required (PA unit and tele-
*4.6.2.6*    phone).

*Compressed air:* None required.

*Practically a duplication of the previous paragraph and also of 30.4 and 30.5.*

*App. 1*    **30.3.** *Steam.* All services to be designed to normal standards.

**30.3.1.** *Condense.* All services to be designed to normal standards.

**30.4.** *Water*

**30.4.1.** *Water, fire.* Existing provisions for the area adequate.

*4.7.4*    **30.4.2.** *Water, raw domestic.* Supply required to room 8.

*The various requirements for water dealt with in the subparagraphs of 30.4.1 to 30.4.9 have already been covered adequately in paragraphs 30.1 and 30.2.*

*App. 2*    **30.4.3.** *Water, raw process.* None required.

*4.6.2.2*    **30.4.4.** *Water, hot domestic.* Local Sadia water heater to be provided in room 8.

*App. 2*    **30.4.5.** *Water, hot process.* None required.

**30.4.6.** *Water, demineralized.* None required.

**30.4.7.** *Water, distilled.* None required.

**30.4.8.** *Water, low conductivity.* None required.

**30.4.9.** *Water, cooling.* None required.

**30.5.** *Gases*

**30.5.1.** *Gas , town.*   None required.
**30.5.2.** *Gas, argon.*   None required.
**30.5.3.** *Gas, argon/nitrogen.*   None required.
**30.5.4.** *Gas, oxygen.*   None required.
**30.5.5.** *Gas, acetylene.*   None required.

*4.7.1*   **30.6.** *Electricity*
**30.6.1.** *Normal supply.*   To meet requirements of para 25.1.1.

*App. 2*   **30.6.2.** *Essential supply.*   None required.
*App. 1*   **30.6.3.** *Street lighting.*   No change.
*App. 2*   **30.6.4.** *Trace heating.*   None required.
**30.6.5.** *Signal and control cables.*   Not applicable.
**30.7.** *Drains*
**30.7.1.** *Toxic material effluent.*   To meet the requirements of paragraph 30.2.
*Paragraph 30.2 does not contain any requirement for toxic material effluent, only for hand wash water.*
**30.7.2.** *Trade waste.*   None required.
**30.7.3.** *Sewage.   None required.*
*Not "foul drain" this time. See paragraph 29.1.*
**30.7.4.** *Storm water.*   None required.
**30.8.** *Telephones.*   To meet requirements of paragraph 30.2.

*App. 2*   **30.9.** *Compressed air.*   None required.
*4.2*   **30.10.** *Roads and paths.*   To meet requirements of paragraph 30.1.
**31.** *Furniture.*

*4.6.2.4,*   **31.1.** *Miscellaneous*
*4.6.1*

One clock, wall type, in room 8
*If the deduction (see comment against paragraph 26.4) from paragraphs 3.3 and 12.1.2 is correct, this requirement cannot be right.*
Mobile shoe barriers

*App. 2*   **31.2.** *Furniture.*   None required.
*It seems that there is "furniture" and furniture.*
**31.3.** *Mirrors.*   None required.

*App. 2*   **32.** *Security*
**32.1.** *Special requirements.*   None.

**32.2.** *Security locks.* None required.

**33.** *Plant*

4.8      **33.1.** *Ventilation.* Ventilation provided to meet the requirements of paragraph 22.2.

*Paragraph 22.2 is "special safety requirements" and states that there are none.*

**33.2.** *Industrial safety.* No special provisions required.

*App. 2*      **34.** *Machine tools and fixed equipment.* Not applicable.

*4.14*      **35.** *Testing.* A test team will be formed at the appropriate time. The formation, composition, and terms of reference of this test team will be the responsibility of the area manager.

*App. 2,*      **36.** *Spares, operating handbooks, maintenance man-*
*3.1*      *uals, and record drawings.* No operating handbooks or maintenance manuals are required. Record drawings of civil, electrical, and mechanical services will be provided.

*App. 2*      **37.** *Planning approval and industrial development certificate.* Not required.

*App. 2*      **38.** *Moundings.* Not required.

**39.** *Sign writing and identification*

*4.11*      **39.1.** *General.* Identification requirements of items and services are stated in appropriate specifications and codes of practice. Preprinted adhesive tape and preprinted labels of approved design and manufacture are preferred for identification use.

*Specifications should be precise and not indicate a "preference."*

*4.6.2.8*      **39.2.** *Pressure vessels and lifting apparatus.* The safe working load of 1 metric ton to be painted on both sides of both monorails, together with the safety reference number.

*App. 2*      **40.** *Position of car and motor cycle parks/cycle shelters.* None required.

*Paragraph 15.2 has already said that there is no provision for parking on cycle racks.*

## Detailed, Fully Cross-Referenced Index to Specification Z 100 in Its Original Form

| Item | Paragraphs |
|---|---|
| Access | 2.4 |
| *Access, pipes | 29.8.1 |
| *Access, walkways | 6.2, 27.3 |
| *Accommodation for ancillary staff | 4 |
| *Acetylene gas | 30.5.5 |
| *Acoustic | 9.3 |
| *Air bricks | 9.5 |
| *Air sampling | 23 |
| *Alarms, controls and | 26, 26.1 |
| *Alarms, glove box | 26.2.5 |
| *Ancillary staff | 3.3, 3.4 |
| *Antivegetation barrier | 14 |
| Associated drawings | 5 |
| *Automatic fire detection | 19.1, 19.2 |
| *Argon gas | 30.5.2, 30.5.3 |
| Basins, wash | 29.9 |
| *Battery chargers | 26.2.4 |
| *Bench lighting | 24.4 |
| *Boundary walls | 17.1 |
| *Building alarms | 26.2.6 |
| Building classification | 19.7 |
| Building description | 1 |
| Cabin hooks | 11.10 |
| Capacity, crane | 27.1 |
| *Car parking/cycle racks | 15.2, 40 |
| Ceilings | 10 |
| Change area equipment | 16.1 |
| Clocks, electric | 26.4 |
| *Compressed air | 28, 30.9 |
| *Condensate | 30.3.1 |
| *Conductive floor and screening | 25.10 |

(continued)

| Item | Paragraphs |
|------|-----------|
| Electric alarms and clocks | 11.3 |
| Electrical services, internal | 25 |
| *Electrical substation | 12.5 |
| Electric clocks | 26.4 |
| Electricity | 30.6 |
| Normal | 30.6.1 |
| *Essential | 30.6.2 |
| *Street lighting | 30.6.3 |
| *Trace heating | 30.6.4 |
| *Signal and control cables | 30.6.5 |
| *Emergency exits | 11.9, 19.5 |
| *Emergency lighting | 24.3 |
| *Entrance | 17.2 |
| *Essential supplies | 25.1.4, 25.4.3, 30.6.2 |
| *Evacuation routes | 19.4 |
| Existing door no. 65 | 11.1.1, 11.2.1, 11.3.2.1 |
| External services | 30 |
| *Types (except telephone) | 30.1 |
| Loads | 30.2 |
| *Steam | 30.3, 30.3.1 |
| *Water | 30.4, 30.4.1, 30.4.9 |
| *Gases | 30.5, 30.5.1 to 30.5.5 |
| *Extract ventilation | 21.1.1 |
| *Extra low voltages | 25.5 |
| *Explosives | 2.1 |
| Final circuits | 25.12 |
| Finish, ceilings | 10.1 |
| Finish, walls | 9.2 |
| *Fire alarms | 26.2.3 |
| *Fire and toxic material warning systems | 19 |
| Fire certification | 2.5.1 |
| *Fire control box | 18.3 |
| *Fire evacuation alarms | 19.3 |
| Firefighting appliances | 18 |
| Fire resistance, doors | 11.4 |
| Fitments, special | 16 |

*(continued)*

| Item | Paragraphs |
|---|---|
| *Laboratory peeps | 11.8 |
| *Layout of plant | 19.6 |
| Layout of rooms | 2.3 |
| Lifting appliances | 39.2 |
| Lifting capacity, monorail | 27.1 |
| Lighting | 24 |
| Type and intensity | 24.1 |
| *Special requirements | 24.2 |
| *Emergency | 24.3 |
| *Crane | 24.4 |
| *Bench | 24.5 |
| *Load bearing, roof | 6.3 |
| *Loading, floors | 7.4 |
| *Loads, external services | 30.2 |
| *Steam | |
| *Condense | |
| *Water | |
| *Gas | |
| *Electricity | |
| *Drains | |
| Telephone | |
| *Compressed air | |
| Location of sockets | 25.1.1, 25.6.1 |
| Locks | 11.3.2 |
| Locks, internal | 11.7 |
| *Machine tools, etc. | 34 |
| *Maintenance manuals | 36 |
| *Mess rooms | 26.6 |
| *Mirrors | 31.3 |
| *Miscellaneous cranes, hoists, etc. | 27.4 |
| Miscellaneous furniture | 31.1 |
| Miscellaneous internal electrical services | 25.6 |
| *Moundings | 38 |
| M panels | 25.4.2 |
| New door (No. 70) | 11.5 |

(continued)

| Item | Paragraphs |
| --- | --- |
| *Services needing essential supplies | 25.11.3 |
| *Sewage, foul drain | 29.1, 30.7.3 |
| *Signal and control cables | 30.6.5 |
| Sign writing | 39 |
| Single-phase AC | 25.1.1 |
| Socket locations, electrical | 25.1.1, 25.6.1 |
| Spares, etc. | 36 |
| Special fitments | 16 |
| *Special requirements, lighting | 24.2 |
| Special safety precautions | 2.6 |
| Special safety reqirements | 22.2, 25.11, 27.5 |
| *Stabilized voltage | 25.1.5 |
| *Standby services | 25.3 |
| *Steam | 30.3 |
| *Condense | 30.3.1 |
| *Storage facilities | 17.3 |
| *Storm water | 29.4, 30.7.4 |
| *Street lighting | 30.6.3 |
| Telephones and PA system | 26.5 |
| Telephones | 26.5.1, 30.8 |
| Testing | 35 |
| *Thermal insulation, roof | 6.4 |
| *Timber, floor | 7.6 |
| Toxic material | 2.2 |
| Drains | 29.2 |
| Effluent | 30.7.1 |
| Level of drains | 29.6 |
| Solid waste | 2.7 |
| *Trace heating | 30.6.4 |
| *Trade waste | 29.3 |
| Ventilation | 22, 33.1 |
| Wash basins | 29.9 |
| Water | 30.4 |
| *Fire | 30.4.1 |
| Raw domestic | 30.4.2 |
| *Raw process | 30.4.3 |

(continued)

| Item | Paragraphs |
|------|-----------|
| Hot domestic | 30.4.4 |
| *Hot process | 30.4.5 |
| *Demineralized | 30.4.6 |
| *Distilled | 30.4.7 |
| *Low conductivity | 30.4.8 |
| *Cooling | 30.4.9 |

*Items identified by an asterisk are considered to be items of redundant information.

*Note:* This index contains about 260 entries, over 150 (or about 60%) of which relate to redundant information.

## REVISED VERSION OF SPECIFICATION Z 100*

### Change of Function Specification and Building Modification

**1.** *Purpose of the Specification*
  **1.1.** This specification describes the modifications that are to be made to the building to include a toxic material container maintenance workshop. These modifications include structural changes to rooms 8 and 9 only.
  **1.2.** Only actual new requirements and changes to existing features and services are dealt with in the specification proper. Features and services that are either unchanged or are not required are listed in appendices to the specification.
**2.** *Scope of the Specification*
  **2.1.** The specification describes the various modifications that are required and details the various items of additional equipment and services that will have to be provided.
**3.** *Related Drawings and Other Documents*
  **3.1.** *Drawings*
  Modified building plan
  Modified electrical layout
  Modified mechanical layout
  **3.2.** *Other Documents.*

---

*Prepared according to the principles discussed in this book.

Paint specification
Electric lamp specification

**4.** *Requirements of the Specification*

**4.1.** *Brief Description of the Building and Required New Function*

**4.1.1.** The building is mainly single story with an extract plant room and a tank room at the first-floor level. It has concrete floors and roof and brick walls. There is a ventilation system, with steam background heating and electric lighting.

**4.1.2.** The new function of rooms 8 and 9 is to provide a toxic material container maintenance workshop and accommodation for the required personnel.

**4.1.3.** A new servicing area will be provided in room 8 and in room 9 there will be reception and dispatch areas with an area for personnel washing. Services will be provided to maintain a suitable environment during normal conditions.

**4.2.** *New Works*

**4.2.1.** To provide a new access to the building for personnel and for vehicles delivering and removing containers, an additional concrete hardstanding will be provided. It will give access to the new door at the eastern end of room 8 from the existing roadway.

**4.3.** *Structural Modifications*

**4.3.1.** The existing lintel and brickwork up to ceiling height over the door opening between rooms 8 and 9 are to be removed. The opening then to be made good to existing standards.

**4.3.2.** The window and frame in the east wall of room 8 is to be removed. The lintel, sill, and all brickwork up to ceiling and down to floor level are to be removed. The width of the opening is to be adjusted to suit the new frame, which is to be fitted as detailed in paragraph 4.4.2.

**4.4.** *Doors*

**4.4.1.** Door 5 in room 9 is to have its blanking boards removed and the opening fitted with a new door, timber faced and 50 mm thick.

**4.4.2.** The new opening in the east wall of room 8 is to be fitted with double doors, timber faced and 75 mm thick. Including the new door frame the overall dimensions are; height 2828 mm; width 1970 mm. The center, top, of the doors is to be fitted around the monorail that is to be installed in this room.

**4.4.3.** Both doors are to have half-hour fire resistance.

**4.4.4.** Both doors are to be fitted with locks as follows:

**4.4.4.1.** Door 5 is to be fitted with a mortice lock accessible from the passage only. The lock is to be provided with two keys on separate rings and each with a brass identification disk stamped with the door identification number. This door is also to be fitted with a Briton 600F floor closure unit.

**4.4.4.2.** Door 70 is not to be made self-closing and is to be fitted with a mortice lock with a key "mastered" XLA.

**4.5.** *Floors, Ceilings, and Walls*

**4.5.1.** The floors of the rooms are to be rescreeded to renew the surface and covered in linoleum.

**4.5.2.** The ceilings are to be repainted using chlorinated rubber paint to specification.

**4.5.3.** The walls are to be repainted after the structural alterations, also in chlorinated rubber paint.

**4.6.** *New Fittings and Equipment*

**4.6.1.** *Furniture Items*

**4.6.1.2.** Two mobile shoe barriers
Three sets of five coat hooks
Two overshoe bins

**4.6.2.** *Other Items*

**4.6.2.1.** Two wash basins are to be fitted in room 9 at standard height in the center of the back wall of the room.

**4.6.2.2.** A Sadia electric water heater is to be mounted on the back wall of room 9 so that it can service the wash basins.

**4.6.2.3.** An electric hand drier is to be fitted on the back wall of room 9 adjacent to the two wash basins.

**4.6.2.4.** An electric clock is to fitted near the ceiling of room 8 on the east wall.

**4.6.2.5.** A PA loudspeaker is to be fitted on the north wall in the northwest corner of room 8.

**4.6.2.6.** A telephone is to be fitted in room 8 near to door 70.

**4.6.2.7.** Mains-fed fluorescent lamps complying with the relevant British Standards are to be used throughout. The numbers to be installed are to be such that the light intensity at floor level in room 9 will not be less than 300 lux. In room 8, also at floor level, the light intensity will not be less than 150 lux. The light switches for these lamps are to be grouped on the north (right) side of the opening between the two rooms.

**4.6.2.8.** Two 1-metric-ton-capacity hand-operated monorail hoists are to be installed, one in each room. Each is to have the safe working load painted on both sides of each monorail, together with the safety department reference number.

**4.6.2.9.** 12-mm plates are to be welded at each end of each monorail to act as stops.

**4.7.** *Services*

**4.7.1.** All electricity supplies to be ac single-phase and 240 volts.

**4.7.2.** Two "M" panels, or equivalent, are to be fitted in room 9 on the north and south walls. An earth leakage trip is to be provided in the supply to the DFB, supplying all the socket outlets called for.

**4.7.3.** A total of six 13-ampere socket outlets is to be provided, four in room 8 and two in room 9. An additional 13-ampere switched socket outlet is to be provided in room 9 for the hand drier.

**4.7.4.** The supply for the electric water heater is to be a fixed, permanent supply.

**4.7.5.** Only a domestic cold water supply is required. It is to be connected to the two wash basins and the water heater.

**4.7.6.** The wash basin taps should be of the elbow-operated type and the basins should be fitted with strainers and traps of stainless steel or approved plastic.

**4.7.7.** The outflow from the basins should flow by gravity to the sump tank in the building via the toxic drain. The flow is not expected to exceed 20 gallons per day. Pipes used should be of stainless steel or an approved plastic. The runs should be such that the pipes are self-draining and free from the possibility of blockage and/or air locking.

**4.8.** *Toxic Material Safety Precautions*

**4.8.1.** The approved toxic material containers to be serviced will all have been certified "clean" for toxicity before arriving at the building. However, small amounts of toxicity may be exposed during dismantling and cleaning of the containers.

**4.8.2.** Other than adequate ventilation for which the existing system is satisfactory, special safety precautions are not required.

**4.9.** *Fire Precautions*

**4.9.1.** Firefighting appliances (fire extinguishers) will be specified, and supplied, by the chief fire officer (CFO).

**4.9.2.** Fire certification will be arranged by the CFO by modification to the existing certificate or by a new certificate, as may be appropriate.

**4.10.** *Building Classification*

**4.10.1.** The building classification symbol, and its location, will be determined in consultation with the CFO.

**4.10.2.** The building manager will order and arrange for the fitting of any new symbols.

**4.11.** *General Identification*

**4.11.1.** Identification requirements for individual items and services are as stated in the appropriate specifications and codes of practice. Preprinted adhesive tape and labels of approved design and manufacture should be used for identification purposes.

**4.12.** *Personnel Requirements*

**4.12.1.** The maximum numbers of staff required for the operation of the service are as follows: four industrial and two nonindustrial, all male. Normal numbers of ancillary staff will enter the building as for other buildings in the area.

**4.13.** *Electrical Earthing*

**4.13.1.** All electrical earthing is to conform to British Standards. The main earth connection is that of the incoming power cable to the building.

**4.14.** *Testing*

**4.14.1.** A test team will be formed at the appropriate time and the responsibility for its formation, composition, and terms of reference will rest with the building manager.

## Specification Z 100: Appendix 1

Features, facilities, and the like, which are not changed by the modifications dealt with in the specification are as follows:

Roof construction
Floor construction
Wall finish
Room height
Boundary walls
Damper requirements
Extract ventilation
Storm water drains
Leakage from pipework
Steam services
Water services
Drains services
Street lighting services

Thermal insulation
Floor loading
Air bricks
Antivegetation barriers
Fire control point box
Plenum ventilation
Distribution board location
Accessibility of existing pipework
Self-draining pipework
Condensate services
Electricity services
Fire water services

## Specification Z 100: Appendix 2

Features, facilities, services, and the like, which are not required as a result of the modifications dealt with in the specification are as follows:

Accommodation for ancillary
  services
Rustproofing
Ducts and drainage channels
Electronic alarms
Emergency exits
Car parking and cycle racks
Detectors and alarm systems
$CO_2$ or other permanent
  firefighting equipment
Handrails
Special safety requirements
Special lighting requirements
Crane lighting
Three-phase ac
Essential supplies
Standby services
Essential supplies alarm
  system
Dusttight and flameproof
  installation

Access walkways
Timber
Door glazing
Laboratory peeps
Electrical substation
Special storage facilities
Automatic fire detectors
Fire evacuation alarms
Radiators
Air sampling
Emergency lighting
Bench lighting
Dc supply
Stabilized supplies
Extra low voltages
Overhead busbar system
Special safety requirements
Individual plant isolation
Alarms/indicators
Battery chargers
Intercom systems

Services requiring essential
supplies
Separation of control and
alarm panels
Fire alarms
Glove box alarms
Gases, compressed air
Raw process water
Demineralized water
Low-conductivity water
Trace heating
Sewage drains
Mirrors
Industrial safety provisions
Planning approvals
Explosives
Gases, oxygen
Gases, acetylene

Mess room
Hot process water
Distilled water
Cooling water
Trade waste drains
Furniture
Special security requirements
Security locks
Moundings
Gases, argon/nitrogen
Gases, argon

# 19

# Word Processors and Document Preparation

In organizations that have responsibilities for the preparation of large numbers of specifications, standards, and similar documents, increasing attention is being turned toward the use of word processors and microcomputers as a way to ease and speed up the task of document preparation. The use of word processors is attractive for the considerable storage and sorting facilities that they make available, to say nothing of the extraordinary variety of ways in which text can be manipulated and moved around a document without extensive retyping. But it should be emphasized that the use of word processors for this purpose should be introduced gradually with careful planning and preparation.

Even for small companies the use of a word processor for the preparation of specifications and similar documents can be well worthwhile, especially because of the benefits that will result from the manipulative and text-moving facilities that are available. For example, if a draft document has been prepared using a typewriter and a need arises to insert a new paragraph into an early page of the document, the entire document has to be retyped beginning at the page where the new paragraph has been inserted. If the document is of any significant length, this is a very time consuming exercise. If, instead, one is using a word processor, the situation is completely changed. The new paragraph is inserted where desired and the word processor automatically adjusts that and all the following pages accordingly, without further effort on the part of the writer.

However, when planning to introduce a word processor to take advantage of its storage and sorting facilities for the preparation of documents using prerecorded "standard" requirement paragraphs, it is recommended that a careful plan be prepared and followed. The steps listed below may be taken as a simple guide.

1. Select a system of similar documents.
2. Identify as many "standard" requirement paragraphs as possible from within the document system.
3. Prepare a list of basic requirement groups. (For example, from the case study specification in Chapter 18 the following groups can be identified: Alarms; Drains; Electricity; Gases; Toxic materials; Water; and many others.)
4. Sort the "standard" requirements into the various basic groups.
5. In the various basic groups, sort alphabetically "standard" requirement paragraphs. (For example, again from that case study specification in Chapter 18, under Water one would find: Cooling; Demineralized; Distilled; Fire; Hot domestic; Hot process; Low conductivity; Raw domestic; and Raw process.)
6. Under the various basic group titles, enter into the word processor memory all the "standard"requirement paragraphs. For later convenience they may be given suitable reference numbers.

One now has a basic data bank of "standard" requirement paragraphs from which one can draw for the preparation of new documents. When a new document has to be prepared, it is suggested that the following steps, or a similar set, be followed.

1. For the proposed new document, prepare a list of the basic requirement groups that it is thought will be required to make up the full requirements of the new document, and put them into the desired order for the purpose of the new document.
2. From the word processor data bank, call up in turn each of these basic requirement groups.

3.  As each group begins to appear on the screen, run down it to select from the various "standard" requirement paragraphs in the group that which is considered nearest in content to the actual specified requirement.
4.  As each "standard" requirement paragraph is selected, transfer a copy of it to a spare group in word processor memory.
5.  When "standard" requirement paragraphs have been selected for all the basic groups specified, print them all out as a separate document.

This single document which has been printed out by the word processor is the first draft of the new document. It must now be carefully reviewed to ensure that the "standard" requirement paragraphs selected exactly meet the needs of the new document. It is very likely that some, if not all of them will require modification before they will exactly fit the bill for the new document. It must be emphasized that insofar as "fitting the bill" is concerned, "near enough is not good enough." A reexamination of the examples of typical specification faults (given in Chapters 4, 5, and 6 should show how necessary such an examination is, and also the need for an exact fit in requirements. It must be emphasized that the examination should not neglect grammar and punctuation.

In this context one is reminded of a story said to date from the seventeenth century:"For want of a nail the shoe was lost. For want of a shoe the horse was lost. For want of a horse the battle was lost. For losing the battle the rider's head was lost!"

However, there is a much more salutory example of the type of problem that can be caused by the absence of a simple comma. In the last months of 1987 the high court in the United Kingdom discovered that they had a very messy legal problem on their hands. This resulted from the omission of a comma in a critical place in an act of Parliament that had come into force in about 1984. This act concerned procedures for divorce actions. What the missing comma had done was, effectively, to cancel thousands of divorces that had been obtained under that act of Parliament. Many of those who had thought they were divorced but were not, had remarried and been informed that they are, legally, bigamists.

It is not very often that such catastrophic errors occur, but sometimes they do. It is also true that a comparable error on a specification or other document is not likely to have such serious consequences. However, that kind of error could be embarrassing and costly for companies (more LOST COSTS). There are plenty of examples in Chapters 4, 5, and 6.

In concluding this brief dissertation on the use of word processors for the preparation of specifications, standards, and similar documents, it is desirable that the following points be emphasized. Use of this type of equipment can be most helpful and time (and cost) saving. However, care must be exercised to ensure that "standard" requirement paragraphs are carefully checked to ensure that they exactly fit the context of the new document.

Not to carry out such a check could cause serious problems later to the eventual users of the document. There are plenty of examples in Chapters 4, 5, and 6 to show just how that can happen.

# 20

# Conclusions

This book has described happenings that over a period of many years convinced me that specification writing presented a major problem and helped me to determine how the problem could be overcome. It can be overcome by the adoption of the concept of specification management. In its entirety this solution does look fairly simple. However, its development was not quite as simple as had been thought! It was arrived at gradually as the various procedures described in Chapters 8, 10, 12, 13, and 14 were developed and proved their worth. That was the first stage.

The second stage was the development of a two-day short course which has been successfully presented in a number of countries around the world. Naturally, improvements were made in the course as time went by and suggestions were received from participants. It was as a result of these comments that the next stage, this book, was born.

With its publication, the concept and its principles can be made available to a much wider audience around the world. It was also clear from many comments made by course participants from many countries that these specification problems had not been recognized until they had been highlighted by the course. Then they became almost blindingly obvious. A common remark was: "How could I not have seen that for myself?" The answer was, of course, the old one of "not seeing the wood for the trees."

The final comments to be made should deal with those LOST COSTS. The fact of their generally unrecognized existence cannot be reiterated sufficiently. The LOST COSTS should *never* be forgotten. It may be thought that considerable costs will be involved in introducing specification management. Costs there will be, but they will be nowhere as great as might have been thought. Certainly not as much as the fire insurance premium costs mentioned in an earlier chapter, but the overall savings can be very much greater.

Almost all of the costs that result, other than the overhead costs taken up by reading and comprehension times, from the use of specifications are credited to other accounts, while specifications get off scot free. If the principles in this book are put into practice, those other costs will be reduced significantly as well as the overhead costs, even if they may not be credited to the use of new and better quality specifications. But the fact will be known by those who work with specifications. Because the quality of specifications is improved, LOST COSTS are reduced. That is the justification for it all: the LOST COSTS.

# Index

Printed in the United States
by Baker & Taylor Publisher Services